A Field Guide to Geophysics in Archaeology

John Oswin

A Field Guide to Geophysics in Archaeology

Springer

Published in association with
Praxis Publishing
Chichester, UK

PRAXIS

Dr John Oswin
Geophysics Team Leader
Bath and Camerton Archaeological Society
Bath
UK

Front cover photographs courtesy Bath and Camerton Archaeological Society.
Rear cover photograph courtesy English Heritage.

SPRINGER-PRAXIS BOOKS IN GEOPHYSICAL SCIENCES
SUBJECT *ADVISORY EDITOR*: Philippe Blondel, C.Geol., FGS., Ph.D., M.Sc., Senior Scientist, Department of Physics, University of Bath, UK

ISBN 978-3-540-76691-9 Springer Berlin Heidelberg New York

Springer is a part of Springer Science + Business Media (*springer.com*)

Library of Congress Control Number: 2009925774

Cover design: Jim Wilkie
Project editor: Rachael Wilkie
Page design and typesetting: David Peduzzi

Printed in Germany on acid-free paper

Contents

List of Illustrations

Figure 5.13 There are two ways to hold the mag, either across your body or parallel to it. When you turn round, turn yourself round, but keep the mag pointing the same way. (a) Head north with mag across you, pointing west. (b) Head south, with the mag still pointing west (change hands). OR: (c) Head north with the mag pointing north. (d) Head south, with the mag still pointing north.

Figure 5.14 For mag without strings, put flags at the top, lines 1, 5, 9, 13, 17 and pegs at the bottom, lines 2, 6, 10, 14, 18.

Figure 5.15 For using a double mag, 1m spacing, put flags at the top at 1m, 5m, 9m, 13m, 17m and pegs at the bottom, 4m, 8m, 12m, 16m, (20m).

Figure 6.1 A very small area surveyed here showed the goal, the shadow of a building protruding from under a church, but did not show if there was anything else in the churchyard.

Figure 6.2 The magnetometer survey was sufficiently large to find the Indian agency it was seeking, and also to pick up signs of Indian settlement and irrigation ditches.

Figure 6.3 The survey in Shetland is mounted on a map as it comprised separate surveys east and west of the burn. There is intense activity in the south-east, and plenty of fainter activity elsewhere. The burnt mounds along the burn show strongly.

Figure 6.4 The picture bottom right is magnetometry. The other five are depth slices obtained by resistance profiling, and are Roman military structures at Satala, Turkey. Depth (a) is 0.125m, (b) 0.51m, (c) 1.09m, (d) 1.77m and (e) 3.4m. Main detail is in (a) and (b), but there is still significant detail in (c) and vestigial effects in (d) and (e).

Figure 6.5 Magnetometer survey of three fields, total area 9ha. An enclosure with internal features is clearly visible in the western field near the kink in the hedgeline.

Figure 6.6 Resistance survey of the area. Some modern drainage lines are clearly evident, but a large building can be seen by the kink in the hedgeline, and a range of buildings below it.

Figure 6.7 Detail of the resistance survey, showing a building some 120m south-west of the main building which is too faint to show on the main plot.

Colour Section

Plate 5 Surveying using a Geoscan FM256 gradiometer. Note the use of a walking string as a guide.

Plate 6 A set of time slices obtained by radar. Each slice represents a different depth under the surface, so you can see at what depth the archaeology is concentrated. Here it is mainly in slices 100–125 and 125–150 cm.

Plate 7 A typical radar set ready for use. The controls are on the buggy, with the antenna towed behind, and the wheel behind that measures distance.

Plate 8 Connectors on a computer (a) connected, (b) disconnected. From the left, USB connector and memory stick, then (other side of mouse connection) RS232D and download cable, then parallel connector and dongle.

Plate 9 An overlay of a res plot of a Roman villa (red) on a mag plot of a ditched enclosure (green). The building lies right over some ditches. Contours are also overlaid.

Plate 10 The picture bottom right is magnetometry. The other five are depth slices obtained by resistance profiling, and are Roman military structures at Satala, Turkey. Depth (a) is 0.125m, (b) 0.51m, (c) 1.09m, (d) 1.77m and (e) 3.4m. Main detail is in (a) and (b), but there is still significant detail in (c) and vestigial effects in (d) and (e).

Plate 11 Overlay of mag and res (Figures 6.5 and 6.6). Colour is used to distinguish between the instruments. The large building is now plainly sitting over the ditches of the enclosure.

Plate 12 An interpretation sketch of the features shown in Plate 11

Plate 13 Ancillary components you need: (top left) grid corner posts; (bottom left) flags and pegs (30cm pegs); (top right), measuring tapes, 30m and 100m shown here; (bottom right), grid lines made from washing line, with pegs (20cm) and wound on a former, and a rope walking line, wound on a former.

In memory of Gordon Hendy, 1945–2008, late of Upper Row Farm, Hemington, Somerset. It was his enthusiasm for the archaeology on his farm that started the project which led to my involvement in geophysics survey.

Preface

It was not an easy decision to write this book when I was invited to. I had no library or university resources to assist me and help keep ahead of the technology. However, I did have the support of an active and well-equipped amateur society and over the past few years I have had experience in teaching geophysics in practice. Many of my students have come from an arts background, with little appreciation of the science behind the techniques. I therefore decided to write a basic introduction to using the common geophysical instruments in the field. Although I cover the necessary science background, this is essentially a guide to the practicalities of setting up and doing simple geophysics projects. Because it is intended as a guide for use in the field, rather than as a scholarly tome, I use the abbreviations 'res' and 'mag' for the principal instruments. That is very much a practicality when you are out in a field and needing to communicate with each other.

This approach means that I have described proprietary equipment and software in some detail. It should be understood that mention of any proprietary items does not in any way represent recommendation.

Nonetheless, I must thank the proprietors of Geoscan Research Ltd, Bartington Instruments Ltd, TR Systems, DW Consulting and Geoquest Associates for their assistance in preparing this text.

I must also thank Lawrence Conyers of Durham University, Colorado, and Steve N. De Vore of National Parks Survey, both of the United States, for permission to reproduce their data, and also M. Drahor of CNSGAP, University of Eylul, Turkey, for his material reproduced with kind permission of Elsevier Publishing.

From the UK, I must thank Neil Linford of English Heritage for supplying information on caesium magnetometers, and Elaine Jamieson and Graham Brown, also of English Heritage, for their help on GPS equipment. Thanks also to Dr Philip Day of Manchester University for support on ground-penetrating radar.

Other examples are taken from my own work, and I must bear responsibility for any errors.

From the Bath and Camerton Archaeological Society, I must particularly thank Jude Harris for all her work on the illustrations, Keith Turner for his software work and Tracey Williams for reading and commenting on the text. I would also like to thank Owen Dicker, Laurie Scott, Jan Dando, Janet Enoch, Pip Osborne, Debbie Shipp and Frances Liardet for their contributions to the pictures.

Thank you also to Clive Horwood, Publisher and Philippe Blondel, Chief Subject Advisory Editor of Praxis Publishing for urging me to write this book, and their support during its preparation.

1

Introduction

There are books on archaeological geophysics in which the authors describe the science and show the wonderful results that they can get. In this book, I try to be different in that I want to show the wonderful results *you* can get. Most books are written by scientists and assume a scientific background. I write here for those with little background in science, although there is some science which you have to know to understand how to use the instruments. I keep the science as simple as possible, but over-simplifying some concepts can make nonsense.

You will need to know the basics of geophysical survey if you are studying for a degree in archaeology, even if your background is in the arts. You may not use geophysics techniques yourself, but you still need to understand what the geophysics specialists are describing. Your university or college will likely have some geophysics kit which you can use.

You may just be someone interested in the subject after seeing some spectacular results. Many amateur archaeological societies now own or have access to geophysics kit, so there are opportunities for you to try your own survey. You may just want to know how it is done.

Professional geophysics archaeologists may seem to have all the latest equipment and techniques, but they need them in order to be more efficient, to win more contracts, to complete their tasks within time budgets and to give their clients confidence in them.

English Heritage's 2008 reissue of its document *Geophysical Survey in Archaeological Field Evaluation* is primarily aimed at setting standards for these professional operators, and makes no mention of work by students and amateurs. However, these groups have a very important role to play, complementary to that of the professional, and they should not feel put off. The text is daunting to beginners, so this book aims to present the techniques more simply. My teachings are broadly in line with English Heritage, but I believe in easier (but not lax) techniques for beginners.

This complementary role does research for its own sake, in timescales not dictated by the budgets of outside organisations. There are many sites yet to be discovered, many to be surveyed, that will never attract a budget for such work, but they may be surveyed by students under supervision or by competent amateur groups whose time is not restricted. It is, of course, essential that such research is done with the necessary permissions, and it is also essential that the survey is written up properly and the results published in an accessible place. In the UK, this would mean, at least, informing the county archaeological society and the county archaeologist so that results can be added to the Heritage Environment Record (HER). English Heritage is also interested in receiving the results of any geophysical surveys.

There are other groups beside English Heritage, such as the International Society for Archaeological Prospection (ISAP), which is for professionals but admits amateur membership. At time of writing, the Institute of Field Archaeologists (IFA) is in process of setting up a specialist group. You do not have to feel beneath their level if your work is of good enough quality to report.

In all cases, you need to gain confidence that you can use the kit to obtain good results. Surveys do not have to be left to the professionals. There is plenty of scope for the amateur or the student to contribute useful data, and the information must be published, so it must be presentable. Using high technology equipment may seem daunting, but if you can understand the basic principles and develop good operating methods, you will be able to produce good results. In this book, I aim to take you through the basic science, its application to instruments, and how to use them in the field. I will concentrate on the instruments you are most likely to meet: fluxgate magnetometers (Figure 1.1) and resistance meters (Figure 1.2).

There are a number of more sophisticated techniques, such as ground-penetrating radar and caesium magnetometers, and the basic science of these instruments will be covered in the text where appropriate. However, their operation in the field goes beyond the scope of this introductory book, so they will not be covered in detail.

The basic science is given in Chapter 2 and its application to the instruments is covered in Chapter 3. It is possible to operate the instruments without the

Figure 1.1 Fluxgate magnetometers. The two shown here look very different, but are similar in principle, with a detector at either end of a long tube, each measuring magnetic effects at a different height from the ground.

knowledge of their science, so you can skip these chapters if you want to, but if you know how the instruments work, it is easier to understand how best to use them.

I also describe the means and methods of downloading the geophysics data to computers and making interpretations. After that, I provide detailed instructions on how to set up and conduct a survey. Again I have kept to the basic instruments, and also to simple laying out of grids on sites in a way suitable for beginners. Professionals may use larger grids, or even more sophisticated methods of position control, such as Global Positioning

Figure 1.2 Resistance measurement. The frame has a voltage probe and current probe, and there are also remote voltage and current probes at the end of a long cable. Dividing voltage by current gives the resistance.

Systems (GPS), but these are not suitable for newcomers, even though you may progress to such techniques later. I finish with a description of a few sites surveyed with the instruments and methods described in this book, and their interpretation.

Geophysics is just one technique which can be used to aid the archaeologists' understanding by covering ground much more quickly than can be done by excavation. Information about sites can also be gained for instance by map studies, aerial photography and earthwork surveys. They are all complementary to geophysics but are beyond the scope of this book. Best techniques will depend very much on ground conditions. If earthworks do not survive, some of those other techniques will be no use. Equally, there are conditions where geophysics may produce no useful results. Some techniques may only work well in some areas. For instance, resistance techniques may be less useful in districts where there is no good building stone. There is little to be gained in a blanket campaign of survey if there is little expectation of finding features. You may strike lucky but you may waste a huge effort. You need to have an initial plan based on some prior archaeological knowledge of the area. A large blank survey only serves to destroy confidence. One purpose of this book is to help you to gain confidence by understanding just what you can expect of the geophysical methods.

Geophysics is, of course, not just an adjunct of archaeology. Indeed, archaeology is a small-scale specialised use of the geophysical techniques developed to understand the physical origin of the Earth. These are also much used commercially in oil and mineral exploitation. The techniques used in archaeology bear little resemblance to those in larger-scale geophysical operations, and are used for very much smaller areas of survey. Archaeological techniques and instruments may, however, be encountered in forensic work.

Geophysics techniques have been developed over a number of decades, but they have come much more to the fore since the turn of this century as a result of the massive improvement in computer power and data logging. Geophysics surveys generate large quantities of data. Earlier versions of equipment such as magnetometers sometimes even relied on long cables and cords attached to a chart recorder set up on a table in a field. This worked tolerably well, provided nobody tripped over the cord! It also limited the data display to a form not much used now, and which is relatively difficult to read.

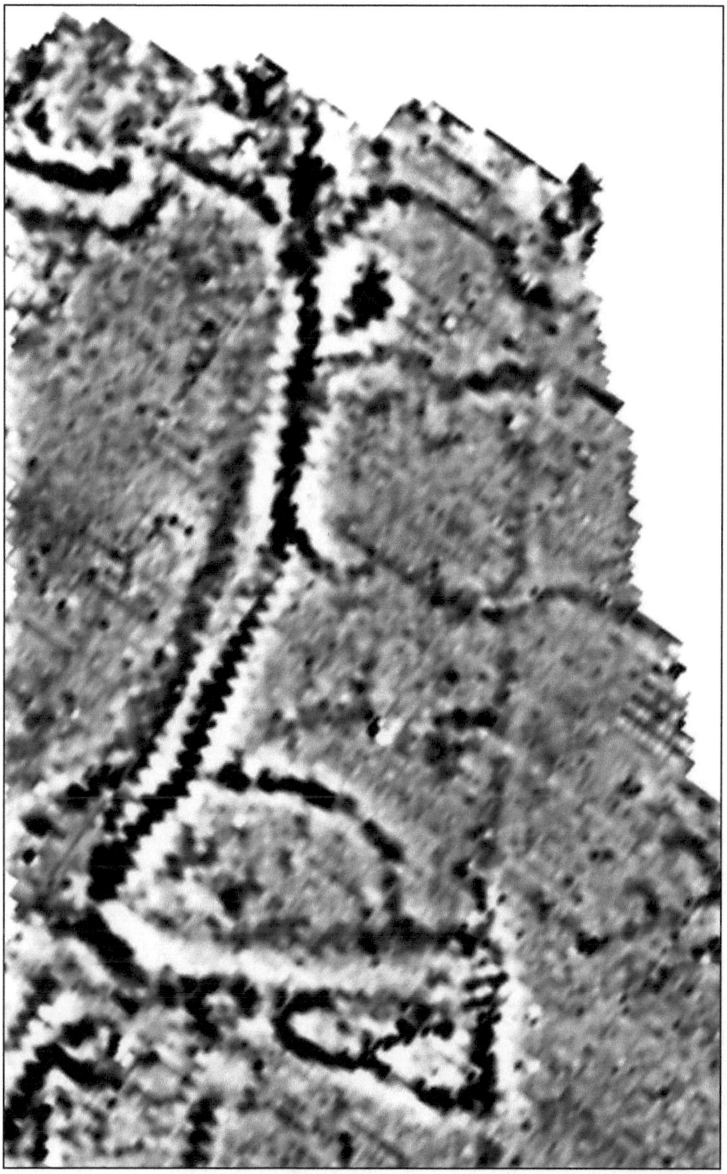

Figure 1.3 Computer output of magnetometer survey. The dark lines represent former ditches in a now-featureless field.

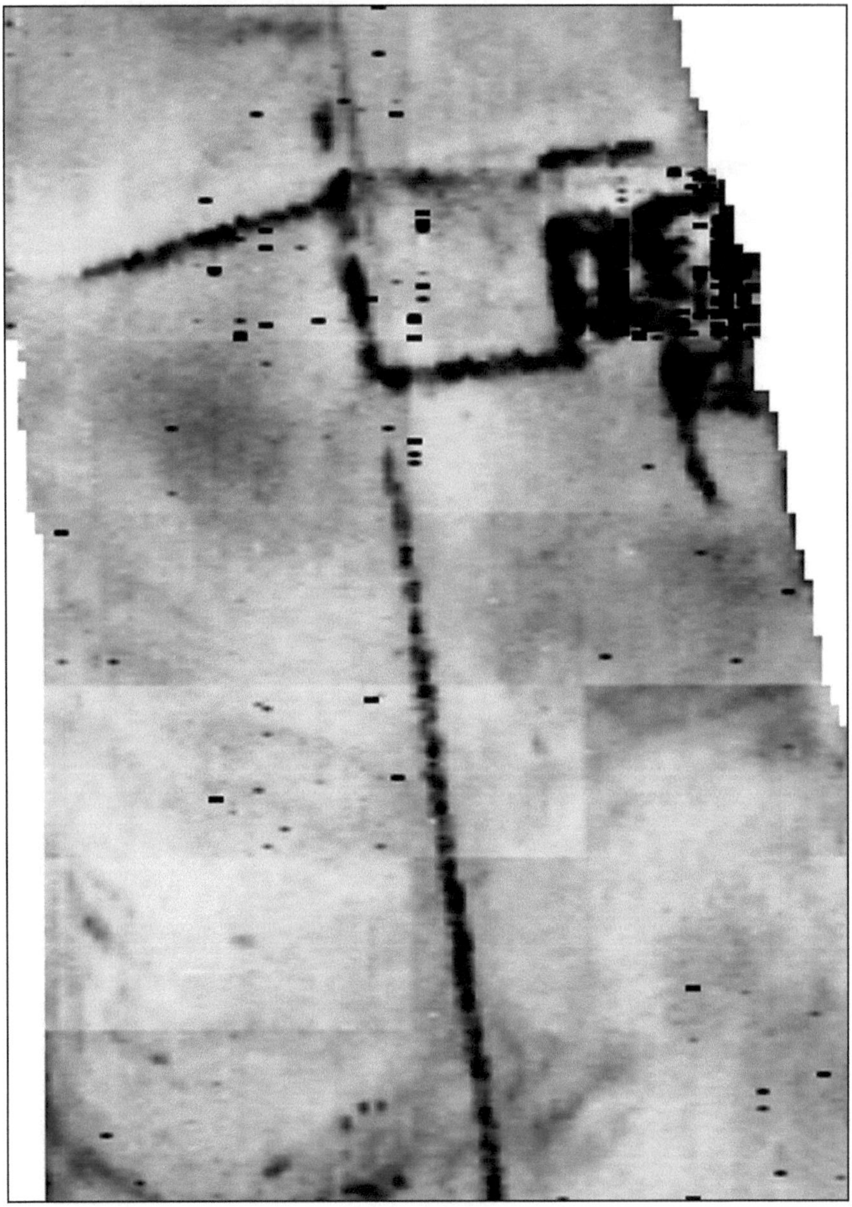

Figure 1.4 Computer output from resistance measurement. The dark lines are stone walls, now buried. There is a building top right.

Current instruments just download to a laptop computer data which can be processed easily and rapidly, although the user must always check that the displayed results make sense. Examples of downloaded magnetometer and resistance data are shown in Figures 1.3 and 1.4 respectively.

One barrier to confidence is the number of long technical words which dog the subject: reactance; inductance; fluxgate; electromagnetic; connector – some I am bound to use, but not more than I have to. Apart from in formal titles and headings, I will abbreviate 'magnetometer' to 'mag', and resistance meter to 'res'. This is particularly important when you are in the field, and need to make instructions as concise and as clear as possible. This book is designed for use in the field.

In order to keep the language simple, I will also refer to the use of these instruments as 'surveying'. The word 'prospecting' is often used in professional literature, but it is less familiar so I avoid it here.

Chapter 2 explains the basic science, and Chapter 3 explains how it is applied to the instruments. Chapter 4 covers downloading to computers. Chapter 5 explains how to use the instruments in the field, and Chapter 6 gives a few examples. Chapter 7 is a summary.

As there is a choice of instruments, I cannot limit descriptions to using one particular device. I have to describe their use in general terms, although I give more detailed descriptions of the devices commonly met in Appendix A. I also apply this to download and processing software in Appendix B.

Appendix C is devoted to the other items you need for a successful survey, such as tapes, lines and pegs. This also covers suitable clothing, both what you need to be outdoors for long periods – sometimes in cold or wet conditions – and what you need to avoid, particularly when using a magnetometer, which is so sensitive, it will detect any magnetic material on you before it detects what is in the ground.

I also include a glossary of technical terms and suggested further reading.

Of course, the best way to become good at geophysics in archaeology is to go out and use the instruments in the field. There is no substitute for experience. However, you have to know the discipline of using the instruments, and unless you have the chance to work with experienced users, you can end up learning the wrong way – or no way at all – to use the instruments. I hope this book will give you a good grounding in geophysics in the field, and give you the confidence and competence to try a survey. If, after a while, you throw it away because it is too primitive, I shall have succeeded in my aim.

2

Science Basics

There are three areas of physics which matter most to our geophysics: electricity, magnetism and light waves. I have used the term 'light waves' as it is much shorter than the correct term, 'electromagnetic waves', but you have to understand that we will be looking at waves well beyond those the eye can see as we must include radio waves, microwaves and infra red. We will look at the basics of each subject in turn. We need to look at the science in this chapter, and then how we might use these effects to detect archaeology in Chapter 3.

2.1 Electricity

Electrical terms – volts and amps – are familiar to us through labels on electrical appliances, but that is not enough when it comes to understanding them, and many people, in confusion, talk about 'volts through' a wire, which is incorrect. Voltage is applied across two points, and then current (amps) flows between those points. Multiplying the volts by the amps tells you the power (watts) that is generated. Dividing the volts by the amps tells you the resistance (ohms) to the current.

Voltage is best thought of as similar to height. If water drops from a height, as in a waterfall, it can be made to work, perhaps turning a water wheel. The greater the height it falls from, the more work it can do. Similarly, the higher the voltage between points, the greater the amount of work which can be extracted.

Current can be thought of as flow. In this case, you cannot only think of the speed the water flows over the waterfall, perhaps one metre per second, but also the cross section area of water flow. A wide, deep stream might have a cross section area (width multiplied by depth) of two square metres, but a small stream might only have an area of a quarter square metre. For these two cases, the volume of water flowing would be two cubic metres per second, and a quarter cubic metre per second respectively. As a cubic metre of water weighs a tonne, that is two tonnes and a quarter tonne each second respectively.

Multiplying the height, say two metres, by the amount of water per second gives the power available, in these cases, four kilowatts for the wide stream, or half a kilowatt for the narrow stream. Note that if we wanted to get four kilowatts from the narrow stream, we would have to multiply the height by eight, to 16 metres.

If we want to know the power from a tonne of water, we divide the height by the flow. In the first case, we get one, in the second case we get 16. These figures are equivalent to resistance.

In the case of electricity, we use similar figures but different terms. It is also easier to think in terms of watts rather than kilowatts. A kilowatt is a thousand watts. Electrical current is like a flow of tiny charged particles, and if our battery has a 'height' of two volts, and there is a current of two amps, that is four watts, and the resistance is one ohm (this is often shown as the Greek letter, omega, Ω). If the current is a quarter amp, the power is a half watt, and the resistance is eight ohms. If for the latter, we increase the 'height' to 16 volts to get four watts, then the resistance becomes 64 ohms. The analogy between electrical volts and amps, and water height and flow is shown in Figure 2.1.

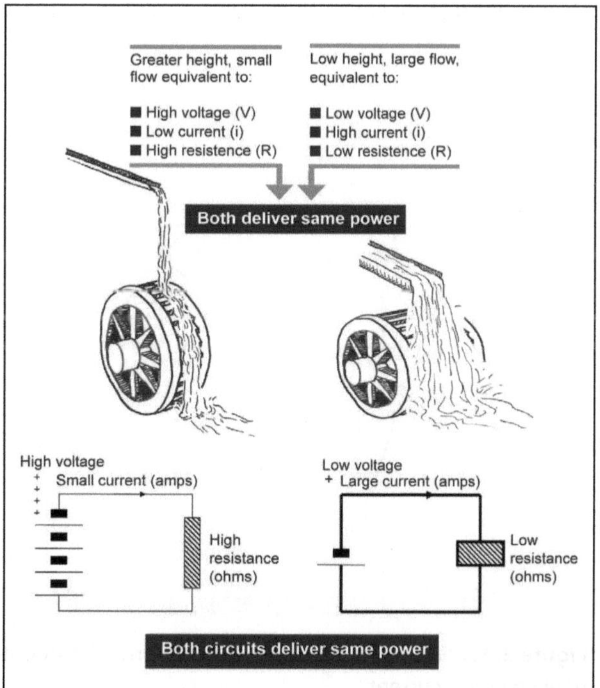

Figure 2.1 Electrical voltage is like water height, and current like water flow. In each case, the 'height' times the 'flow' gives the power delivered, and the 'height' divided by the 'flow' gives the resistance.

You may also come across the term 'resistivity' but that is not strictly the right term in most areas of geophysics. Resistivity is a property of each material which determines the electrical resistance of that material of a particular shape. As normally set up, our instruments measure the electrical resistance of a path through the ground. In one particular case, this may be used to calculate the resistivity of patches of soil.

So far we have talked about voltage as a simple one way flow, from one end of a battery to another. This is known as direct current (DC), but there is another form – alternating current (AC) – where the current flows back and forward, more like a tide than a stream. Batteries produce DC, the mains electricity is AC. DC is simple, but AC has advantages. First we need to relate the alternating current to the direct current in terms of the work it can do.

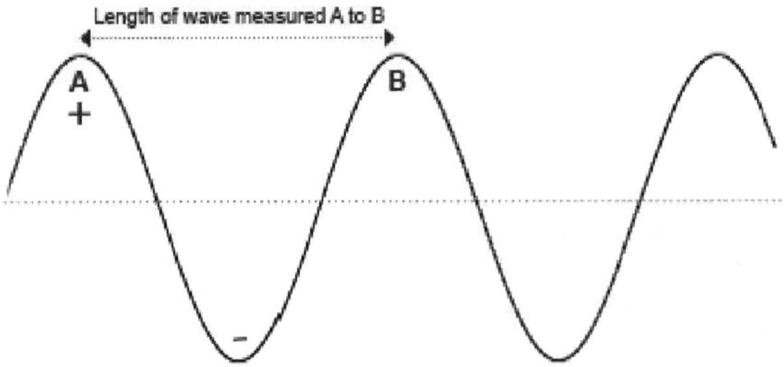

Figure 2.2 A sine wave. This is the curving form of a smooth wave, whether of water or of electrical current.

Alternating current goes from a peak value through zero to a trough – that is, the same value as the peak but in the opposite direction – and then back to the peak, in a smooth curve known as a sine wave (Figure 2.2). The voltage also follows the same pattern. If the power produced (volts times amps) is the same, this must be the average power, midway between the peaks and zero. If we multiplied both volts and amps by a half, we would end up with only quarter the power. To get a value of half, we have to multiply each by the square root of a half, written as $1/\sqrt{2}$ The tick sign means square root, that number which multiplied by itself makes the number you first thought of. It is worth remembering that $\sqrt{2}$is 1.414, and $1/\sqrt{2}$ is 0.707, numbers you will come across regularly.

The voltage or current equivalent to the DC value, is known as the 'root mean square', or RMS value, and is 0.707 of the peak value. Indeed, when you talk about, say 110 volts from the mains, that is the RMS value, and the peak value is 156 volts. From peak to trough is 312 volts.

Alternating current has another property as well as its value, and that is the rate at which it keeps turning back on itself. Depending on where you live, the mains current will go through a cycle from peak to trough and back to peak 50 or 60 times a second. This is a lot faster than the tides at twice a day, but the figure could be a lot higher, up into millions of times a second, and we shall come across these very high figures later. The rate of this cycling is called the frequency and is measured in hertz (Hz), so we refer to mains at 50 Hz. When it reaches a thousand times, we call that a kilohertz (kHz), a million times, we call a megahertz (MHz), and a thousand million times a gigahertz (GHz). It is worth remembering these terms.

AC electrics have a big advantage in that it is very easy to change the voltage from one level to another using a transformer, but when you transform up the volts, you transform down the current in the same ratio. It has another advantage in geophysics in that it stops electrically charged particles building up as the direction of the flow is always changing. We will discuss this later, in Section 3.1.

AC has other properties which behave rather like resistance, but are rather different, and these are called 'inductance' and 'capacitance'. You will need to know the basics of these, but not all the science. Imagine our current and voltage sine waves as in Figure 2.2, but the voltage is at a peak when the current passes zero, and the current peaks when the voltage passes zero, then the average work done will be zero. This is a condition known technically as 'phase quadrature', a term you may come across in specialist books. If the current is ahead of the voltage, this is capacitive. If it lags behind the voltage, it is inductive. These effects pull in opposite directions.

These resistance-like effects are called 'reactance': capacitative and inductive. As the frequency increases, capacitive reactance decreases and inductive reactance increases. There will be one frequency at which the two effects cancel each other out. This is known as the 'resonant frequency', and the electronics will work most effectively at this frequency. This will normally be the operating frequency of the equipment. Knowing about inductive effects is also important, as explained in Section 5.4.

We are used to the idea of an electrical current flowing in a wire, but what happens when a current just flows in the ground? In a wire, it is tightly constrained, but in the ground it spreads out. It must enter the ground at a

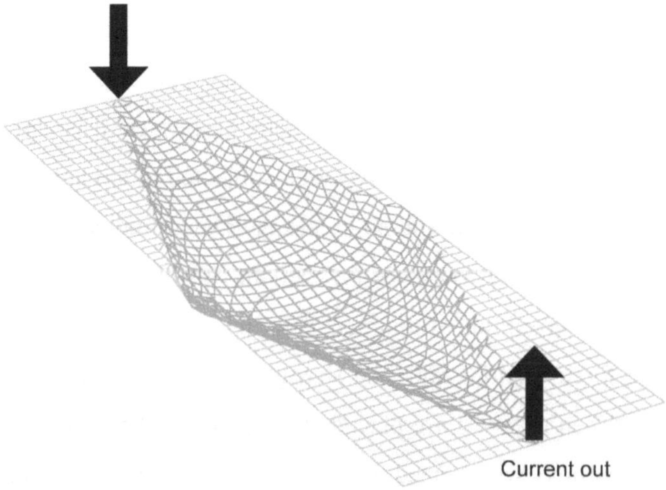

Figure 2.3 Electrical current may enter the ground at one point and leave at another, but between those points it spreads out. It does not travel in a line, as it does in a wire.

point and exit the ground at a point, and at these two points the current is concentrated, but at other points between entry and exit it spreads out, as in Figure 2.3.

The resistance of the ground depends almost entirely on its water content, as electrical current flows quite easily through water. So expect a wet clay soil to have a low resistance (although this will rise as the clay dries out). Sand is dry so will have a high resistance, and rock does not hold water, so tends to be high resistance.

2.2 Magnetism

Magnetism is an atomic property which differs from element to element according to the behaviour of the electrons – the tiniest particles – which orbit in the atom. Iron has particularly strong magnetic properties. It is a property of some types of iron that once they have been subjected to a magnetic field, they can retain that magnetism, and so create their own magnetic field. We just call them magnets. The forces of magnetism can be used to attract or repel other magnets. We think of magnets creating a magnetic field around them, where the forces can be felt. The strength of the field can vary from tiny to massive. When magnets are placed in each other's field, they will try to move to find positions which need the lowest energy to hold them there.

We are used to the idea that the Earth has a magnetic field when we use a compass to find North. The compass is a magnetic detector, but is too crude and insensitive to meet our needs in archaeological geophysics. It works because it is a magnet which can swivel. The compass points in the direction of the Earth's magnetic field, which is north–south. Attracting forces are set up between the north and south poles of magnets, so that the point of the compass is pulled as near to the north pole as it can get. This is the state of lowest energy which the compass needle can find.

The unit by which the 'strength' of the magnetic field (its flux density) is measured is the tesla, abbreviated to 'T'. One tesla is a very large magnetic field, and the Earth's field is about 50,000 nano-tesla (abbreviated to 'nT'). The prefix 'nano' means 'divide by 1000,000,000 (one thousand million)'. The field emanates from the south pole and returns to the north pole. It travels parallel to the surface at the equator and slopes steeply down to the ground

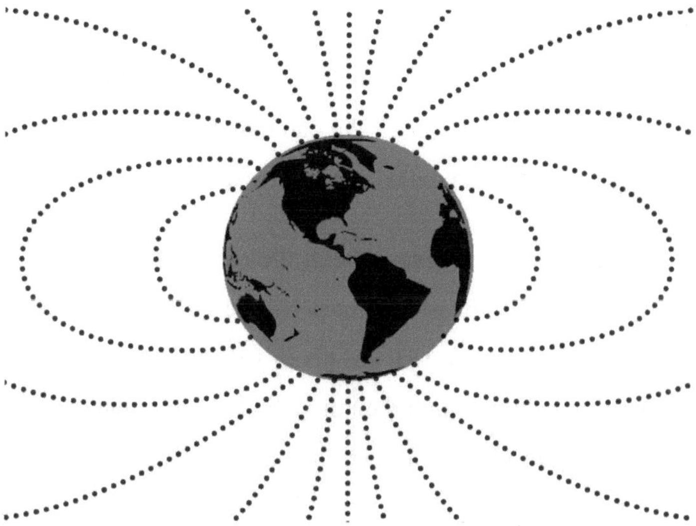

Figure 2.4 The Earth's magnetic field is formed between south and north poles. It is parallel to the surface near the equator and turns steeply into the ground near the poles.

surface in northern parts, such as the UK, and also far south of the equator. This is shown in Figure 2.4.

We might think that the Earth's magnetic field is constant in time and place, but if we could measure it with sufficient sensitivity we would find small variations. Some variation is due to geological effects but there are even smaller, even more local variations. It is these very small variations which benefit archaeologists, but we have to have an idea what causes these changes. We also need to know how to measure them.

We are looking for changes in the Earth's magnetic field of typically 1 to 100 millionths, that is 0.05 to 5 nT. Only very sensitive devices can respond to these tiny amounts. What physics can we use to help us? Metal detectors see magnetic effects, but not this small. We can either detect a change in energy, or we can watch for an effect called magnetic saturation.

If there is a change in the magnetic field, the energies of atoms in that field will no longer be at their lowest and there will be a change to find that lowest state. In the process, energy will be given off, and if we can detect that energy, we will know how much the magnetic field has changed. There have to be enough atoms to give a measurable amount of energy, and it will take a finite time for the energy levels to change and for the released energy to get to the sensors, so this is a relatively slow process. Note that the signal is proportional only to energy in the field, not to its direction.

The first type of magnetometer which relied on this process was the proton magnetometer. It measured the energy of hydrogen atoms in a liquid called methanol. The hydrogen atom is the simplest and smallest there is. The more modern instrument is a caesium magnetometer, which is more sensitive and has a faster response. Caesium is a much heavier atom, but its magnetic behaviour is similar to hydrogen. This allows us another form of energy measurement, and caesium magnetometers are faster and more sensitive than proton magnetometers.

Then there is magnetic saturation. An iron magnet can only be made up to a particular strength, no matter how strong a field it is put into. That is, it saturates. Figure 2.5 shows how a material responds magnetically to a magnetic field. The time to reach saturation can be very short, but it can

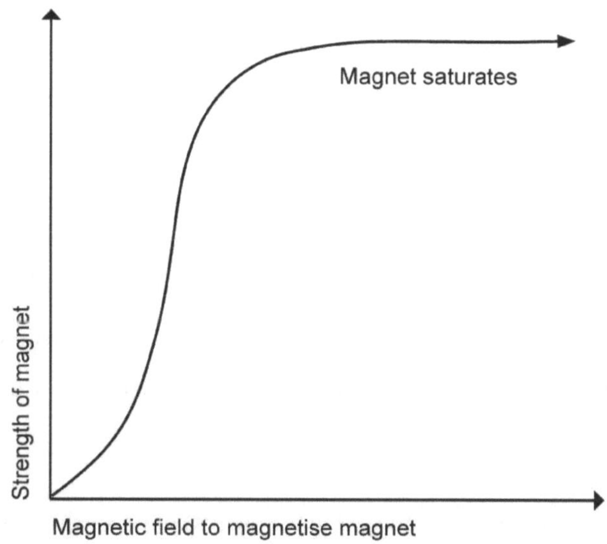

Figure 2.5 Magnetic materials 'saturate'. You can go on increasing the magnetism, but they do not get any more magnetic beyond a certain level.

be measured. The time taken to reach saturation will depend on the level of magnetic field to start with. Even small changes in field can affect the saturation time in a measurable way.

A tube of material called mu-metal (which behaves like iron only more extreme) will not normally allow a magnetic field inside it because of its magnetic properties. However, when it is saturated, it will allow the field in. As it is flipped back and forth between normal and saturated states, the magnetic field – in this case the Earth's magnetic field – is alternately pulled in and driven out of the tube, and the changing field can be detected as an electrical signal in a coil of wire. That signal can be measured very accurately, so even a tiny change in the starting magnetic field can be measured very accurately and very quickly. This provides another means of measurement.

Magnetic fields are often talked of in terms of their 'flux'. The tube acts like a gate opening and closing to the field, and this type of detector is thus known as a 'fluxgate'. Note that it is only sensitive to that portion of the field passing through the tube, so a fluxgate is directional, and is not sensitive to the whole of the Earth's magnetic field. The implications of this will be discussed in Section 3.2.

We have a magnetic field and we have a means of measuring it very precisely. All we need is something in the ground which can give rise to these small changes. Fortunately, there is something in most, but not all soils. It goes back again to iron. There are two types of effect which we see.

Iron is one of the commonest materials on Earth. It exists in various forms depending on which other elements are present. Most of these have some magnetic effect, even if it is only small, but we are now looking for very small changes. Many soils contain high levels of iron. Soils and rocks rich in iron tend to have a yellow or red colour. The subsoil will often have a different iron content from the topsoil. If a hole is dug into the subsoil, it refills over time mainly with topsoil which is washed into it. Thus, a patch in the subsoil looks magnetically like topsoil, and a magnetometer can detect this. If the hole is filled with rotting material, perhaps bits of carcasses and animal waste, this may also be rich in iron and produce a different magnetic signal from the surrounding subsoil. It can even allow growth of microbes

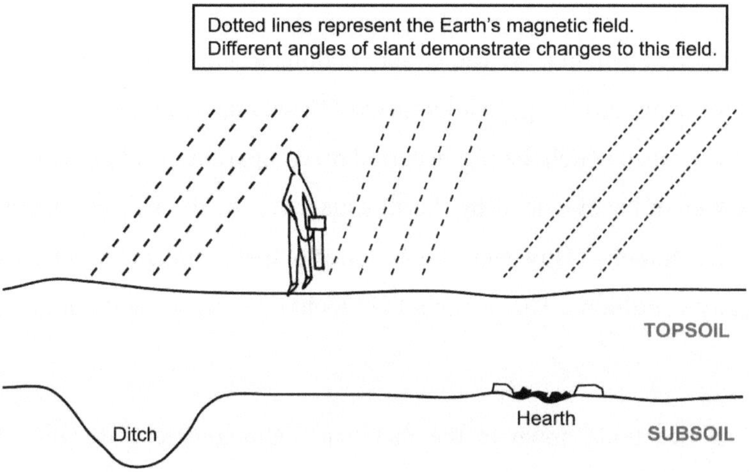

Figure 2.6 The presence of magnetic material in a now-filled ditch will cause a small disturbance to the Earth's magnetic field. Burning will also cause materials to disturb the magnetic field.

which are slightly magnetic. The effect on the Earth's magnetic field is shown in Figure 2.6.

The second effect is from fire. You can destroy a magnet by getting it very hot. Intense heat agitates the magnetic structure to the point where it loses all orderliness and becomes random. When the magnet cools down, the Earth's magnetic field will re-magnetise it, but only weakly. An iron-rich soil will give up its magnetic properties when it gets very hot and regain magnetic properties when it cools. However, the Earth's magnetic field is changing in strength and direction very slightly all the time. The magnetic properties taken back in by the soil (or stone) will be those of the day it cooled, not those it had previously. The magnetometer can detect that change in properties.

There is an added benefit if we can be sure nothing has moved since it cooled. If we can measure the change in alignment of the magnetic properties, we can estimate the year in which the fire occurred by comparing the alignment with curves of known behaviour of the Earth's magnetic field. This is known as archaeomagnetometry. However, this requires a special magnetometer used in laboratory conditions. This is not a field technique, so we will not discuss it further.

All this science boils down to the fact that magnetometers are good for finding pits and ditches dug into the subsoil and also for finding signs of burning. However, they do not give results on all soils. There has to be a good iron content.

2.3 Light waves

I have used the word 'light' but we need to look wider than that, and consider a whole range of light-like waves which the eye cannot detect. The correct generic term for all of these is 'electromagnetic waves'. These include infra-red, microwaves and radio waves. These have similar physics to light, but lower frequencies. There are also higher frequency waves, such as ultra violet and X rays, but we need not discuss them here. The wave-like motion is similar to the sine wave of Figure 2.2. In Section 2.1, we discussed frequencies, how often you get from one peak to the next in a second, but there is another effect. That is the distance between one peak and the next, known as the wavelength. For any wave type travelling at the same speed, the wavelength decreases as the frequency increases. Indeed, if you multiply the wavelength by the frequency, the result is the speed of the wave.

Earlier in this chapter, we found electric currents making magnetic fields and changing magnetic fields being detected electrically. The two are linked, and light waves use both magnetic and electric fields to transmit energy. Light is one part of the range, or spectrum, of the waves; it is the bit that our eyes can see, but it is only a part.

An electric field is like a voltage across two points, but with no current flowing between them. The field is larger for high voltages and also larger for the smaller the distance between those points. When an electric field changes, it sets up a magnetic field at right angles to itself. And, of course, when the magnetic field changes, it sets up an electric field at right angles to itself. We saw this effect in inductors when we discussed reactance. So we have a self-perpetuating wave of falling electric field, growing magnetic field, falling magnetic field, growing electric field and so on.

This wave travels through space at a constant speed, the speed of light, which is 300,000 kilometres per second, a speed so high as to make everything seem simultaneous, but that is not quite the case as we shall see later. Light is slowed down as it passes through water or glass and is stopped altogether by metals.

Figure 2.7 shows the range, or spectrum, of wavelengths that we may normally meet for light waves in archaeology. It is more usual to quote wavelengths than frequencies with light-like waves. The spectrum extends beyond both of the ends shown here, but the higher end takes us into specialist areas of

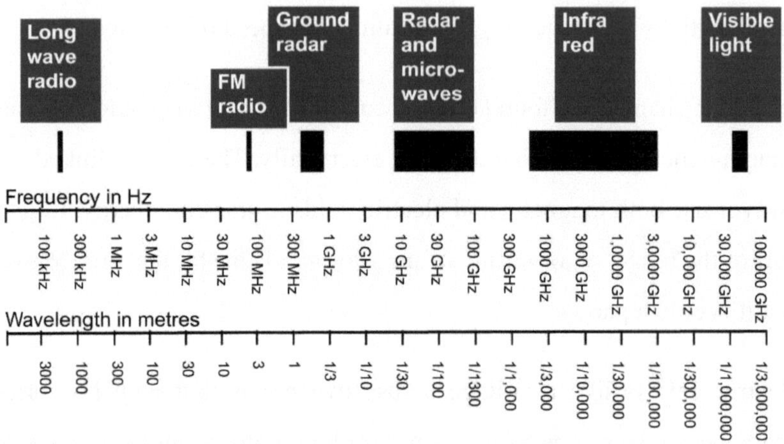

Figure 2.7 Light is only a small part of the spectrum of electromagnetic waves, which includes radio, microwave and infra red. By spectrum, I mean the range of wavelengths.

science. The lower frequencies (very long wavelengths) extend down to the audio range, but are difficult to transmit.

The lowest frequency here (200 kHz) has a wavelength of 1500 m, which is the UK BBC Radio 4 Long Wave transmission. This travels hundreds of miles, over hill and dale, and can be picked up just about anywhere in Europe. Next shown is the band used for FM radio, about 3 m. This is one area usually characterised by frequency, and we are used to hearing about '96.7 megahertz' (mega means a million times). If you multiply that by the wavelength close to 3, you arrive at 300,000,000, which is our speed of light. By the time the wavelength is this short, the wave will not travel very far, so there is less likelihood of interference from another distant radio transmitter, but reception can be badly broken by hills, or even by houses, in the way. However, if the signal comes from the sky, for instance from a satellite, there will be less to get in the way of the signal. Radio waves of various wavelengths can also penetrate the ground, and this may be useful to geophysics.

Check the scale of the graph. It does not move by regular points, but each mark represents the wavelength being divided by 10. This is known in mathematics as a logarithmic scale, and is a useful way of compressing a huge range on to a manageable graph.

Continuing along the scale to shorter wavelengths, we encounter microwaves, or radar around 1 metre to 1 millimetre. The case of ground radar, used in archaeology, will be considered in more detail later. Further along we find infra red (IR) and then at about a half of a millionth of a metre, we see light. Continuing beyond that there is ultra violet (UV), and then X rays with very small wavelengths.

We are used to light from the sun and from light bulbs. It comes in over a range of wavelengths (red has almost twice the wavelength of violet) and in a continuous and quite random manner. If we want more ordered light waves, we need a laser. To see the difference between these two, think of two different ways of walking. You can imagine a group of walkers on a ramble. They all move at about the same speed to arrive at the end together, but they spread out randomly and stop now and then, in a rather disorderly way. This is very different from an army battalion, which marches with fixed pace at a fixed rate so they all stay close together in formation. If you know the length of their pace, you can even calculate their speed by timing their footfall.

A laser can create this highly ordered form of light wave, and can create it in very short bursts so it is more useful for measuring. If you know that there are only 10 wavelengths of light at a half of a millionth of a metre, then that pulse will only be five millionths of a metre long, so you can measure to that precision. Each type of laser operates only at one particular wavelength, and cannot produce other wavelengths. Outside the region of light, other types of device can create highly ordered wave trains, for instance in the radar band.

We have only talked about the wave travelling through space, although travelling through air is not much different. However, things are a bit different when a wave hits something solid or liquid. Let us start with the effect of water. We noted earlier in this chapter that water can conduct electricity, although not nearly as much as metal can. Still, there is enough conduction to give us readings with a resistance meter. Back at the start of this section, we talked about an electric field set up with no current flowing and altering that to create a magnetic field.

If some current can flow, there will be two bad effects for the light. It will slow down, and it will lose energy, so eventually it will sink to nothing. The energy is turned into heat. We are used to water slowing the wave down, an effect known as refraction, because we can see it. When we put a stick into some water, it appears to bend. This is a result of water having a slower speed of light, about one quarter less than the speed in space. The speed of light through the ground is much less than through air, so the wavelength of radar in the ground is much less than it is in air for any frequency. We need a lower starting frequency in air to retain the wavelengths we want in the ground.

We are not so used to the idea of light heating water, but we are certainly used to radar heating water. This is the basis of the microwave oven, which turns all its radar wave energy into heat.

Mainly, the distance it takes a wave to be reduced to nothing and turned into heat depends on the wavelength, so longer waves persist where shorter waves are wiped out. There are some special exceptions to this rule, however, due to effects at atomic level. For instance, light very close to half of a millionth of a metre (blue-green to our eyes) will travel through water better. This is why there is a green tinge to the light under the sea.

What happens when light bumps into something solid? If it is metal and conducts electricity well, the light wave cannot enter it, but is reflected back. We see this every day as a reflection from a mirror. If the object is transparent like glass, most of the light will go through, although its speed is reduced as it crosses through, and a small amount is reflected. If it is black, the light is just absorbed by the material and none is reflected. Most materials absorb some light and reflect some.

That is what happens on the everyday scale, but it is not quite the same at microscopic scale. Many materials look and feel smooth, but if you check them under a microscope, their surfaces are quite rough. If the surface's spikes and dips are separated by about a wavelength of the incoming light (or radar), it is reflected off in all sorts of directions: it is scattered. This can be a nuisance if too much of the light gets scattered back into your eye (this is what happens when you shine a light in fog), or into whatever you are using as a detector, but it can also tell you that there is something there to be found.

If you are using ordered waves, for instance from a laser, it can be easier to sort out the reflections from the scattering.

We have seen in this section that light is only part of a much wider spectrum of electromagnetic waves, which includes radio and radar. We have looked at some of the basic properties of these waves, such as speed and wavelength, and seen how they may be spoiled by bending, absorption or by scattering.

This concludes Chapter 2. In Chapter 3, we will be looking at the application of this basic science to the different types of geophysics instrument, resistance meters and magnetometers, as well as the applications of light waves of various wavelengths to ground-penetrating radar GPS and Total Stations.

3

The Instruments

In Chapter 2, we described both the science that gives the soil properties to measure and the means by which the instruments make their measurements. In this chapter we will look at the application of this science to real instruments. New instruments come out from time to time, so it is best not to go into too fine detail, lest it be obsolete by the time you come to read about it. In all cases, each machine will have its own operating method which you will have to learn if you want to use it. Details of some of the instruments commonly used are given in Appendix A.

However, it is important to see the science being turned into practical use, and to realise what it takes to make a real instrument fit for use. There are two forms of instrument: 'active' and 'passive'. Active instruments send out energy and detect the returned energy; passive instruments just sense the environment around them – and that can be the environment created by you so you may need to use these with extra care.

3.1 Electrical resistance instruments

In Chapter 2, we recognised that there was a relationship between the voltage across a couple of points and the current that flowed between them as a result of that voltage. This is called the resistance. We also met 'resistivity'. Every material has a resistivity of its own, and if you keep the dimensions of materials the same (same length, same cross section), this unique 'resistivity' can be compared between one material and another by measurement of electrical resistance. In our case, two materials being compared might be earth and stone. Stone has less moisture in it than earth, so it has a higher electrical resistance. If our measurement of resistance changes from one point to the next, then we might conclude that one point has just earth under it, while the other has stone under it. If a number of points which have stone under them line up, we might conclude that there is a buried wall, and that would be useful to archaeology.

First we have to introduce an electrical current into a piece of ground by applying a voltage across it, and that is not quite as easy as it sounds. We could just put two metal probes into the ground, connect a battery across them, measure the voltage across the probes and the current flowing out of the battery, divide voltage by current and calculate resistance, as in Figure 2.1. We could even estimate the resistivity of the ground, but we would most likely get the wrong answers for both, for more than one reason.

One reason, which we have mentioned before, concerns a battery-like effect. There are electrically charged particles in the ground, particularly in the water in the ground, and it is these that let the current flow. If we put probes in the ground, one connected to the plus terminal of a battery and one connected to the minus terminal, 'minus' charged particles will collected around the plus probe and 'plus' charges will collect around the minus probe. These will form their own battery which will try to send current in the opposite direction, so that we measure the wrong current. We can overcome this by replacing our DC battery with an AC source. This will make the plus and minus probes swap over faster than the particles can collect around them, so there is no build-up of charge.

The second reason concerns the probes. We have assumed that they are perfect, and that there is no resistance between them and the ground, but this is not so. Figure 3.1 shows how this affects the measurement.

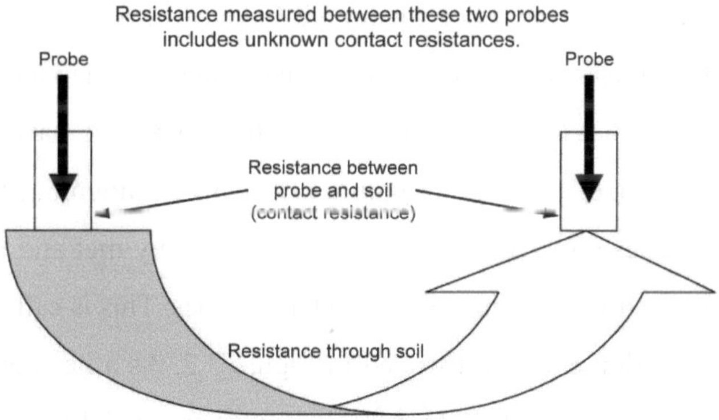

Figure 3.1 You cannot measure ground resistance directly between two probes, because you do not know how much is caused by contact between probe and ground.

Getting current from a metal probe into the ground is quite difficult. There is a 'contact resistance' to overcome. In a simple measurement such as that described above, we would not be able to distinguish between earth resistance and contact resistance. If we want to know the earth resistance, we have to get the contact resistance out of the equation, and there is a way of doing this. We use two probes to measure the voltage and two other probes to measure the current, so each measurement should in theory see the same contact resistance effects so that they cancel out. In science, this is often called a resistance bridge. Fortunately, this works well in practice too, although in some very dry soils there may be discrepancies from probe to probe, and this will give a wrong reading. One of the skills of resistance measurement is recognising when that happens.

We now have four probes, and we have to arrange them so that they give a measurement and so that we can easily move them from point to point. This is most easily done by mounting them on a frame, which can be pushed into the ground for a measurement and then extracted and moved to the next point. If we mount the measuring equipment onto the frame, we have a portable measurement device. There are a number of ways of mounting the voltage and current probes on to a frame. The most commonly met and easily used method uses a two-probe frame and two static probes. This is called the 'twin probe' configuration and is illustrated in Figure 3.2. As a newcomer, this is the method you are most likely to meet. It may seem unwieldy at first having to trail a cable to distant probes, but it makes the frame lighter and easier to push into the ground, and also gives results which are easier to interpret. When I just mention 'res', that will refer to a twin-probe device.

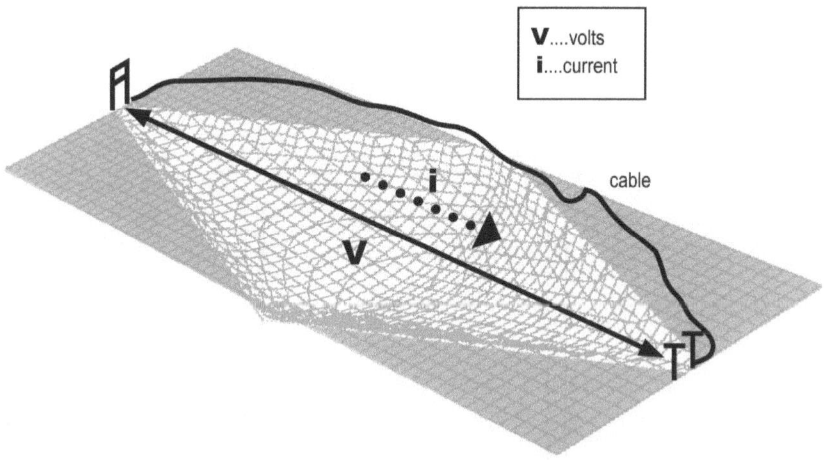

Figure 3.2 The twin probe arrangement actually uses four probes. It measures the voltage between two and the current through the other two. The current spreads out while it travels through the ground.

You may also come across frames with all four probes mounted at equal spacing. It is most likely that these will be arranged with the outer probes passing the current and the inner probes measuring the voltage. This is known as the 'Wenner' configuration, and is shown in Figure 3.3. People have also tried mounting the probes as spikes on the wheels of a cart, to form a 'square' array, front wheels for current, back wheels for voltage. It sounds easier to tow the device like this rather than carrying it, but it is not so easy in practice, particularly through long grass or over rough ground.

In our calculation of resistance, we assumed that the voltage and the current were measured over the same piece of ground. In some probe configurations,

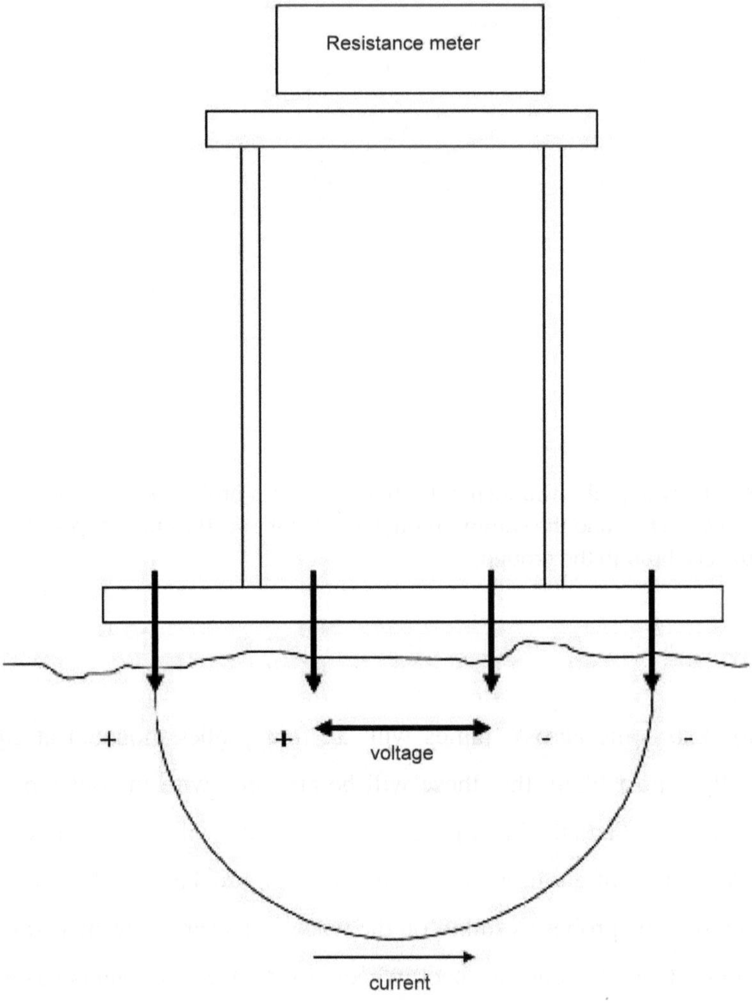

Figure 3.3 Arrangement of probes in the Wenner configuration.

this is not so. This does not matter if the ground is very even in consistency, but in archaeology, this is exactly what we are not looking for! These methods do show a consistent change in resistance as the probes go over a stone, but that is not a simple change, so it is more difficult to understand. Note that the probes do not have to touch the stone, but the stone needs to be beneath the probes.

With the twin-probe arrangement, the frame has a voltage and a current probe, and the distant probes are also one voltage, one current. They are connected back to the frame by a length of cable. The cable has to be low enough resistance that it does not affect the measurements. Providing the frame and the remote pair are far enough away, the measurements appear to be made at the same points relative to their separation. As a simple rule, the separation between frame and remote probes should be at least 30 times the space between the probes on the frame. The error is then 1/30, or about 3% and this has been found acceptable in practice. If the frame has half-metre spacing between probes, the remote probes must be at least 15 metres from the frame. If the frame has one metre spacing, the remote probes need to be at least 30 metres away.

As the separation between the probes increases, the stones that affect the resistance can be more deeply buried, so the device becomes more capable, but moving a one-metre frame is more difficult than a half-metre frame, and the interconnecting cable has to be longer (and that also means fatter if the cable resistance is to be kept down). There could also be a problem if there is bedrock close to the surface, as this may swamp any signal from archaeology above it. Half-metre frames are usually seen as a good compromise.

Figure 3.2 shows the general working arrangement of the twin-probe resistance meter. Current enters the ground through one of the probes on the frame and is collected from the ground at one of the remote probes and returned to the frame by the cable. The voltage from the other frame probe to the other remote probe is also measured by the meter. Providing the probes of each pair are close enough together, compared to the separation between frame and remote probes, each of the pairs can be regarded as being a single point.

You might think that this arrangement would detect every stone between frame and remote probes, producing a cluttered meaningless result, but reality helps us in this case. When electrical current flows through the ground it does not travel straight from frame to remote probe. As the current enters the ground, it spreads out, both to the sides and down, as shown in Figure 3.2. The current becomes concentrated again around the collecting probe. The only places where there is sufficient concentration of current are right next to the frame probes and right next to the remote probes. Once the remote probes are set in the ground, the only place where stones can affect the reading from place to place is immediately under the frame.

Figure 3.4 shows a typical twin-probe resistance meter in use (Figure 3.4 is reproduced as Plate 1 in the colour section for enhanced interpretation). There are two makes in common use, the Geoscan RM15 and the TR/CIA meter. They look very similar. The Geoscan device is more sophisticated, but the TR/CIA was deliberately designed as a low-cost meter. The Geoscan can be configured to other probe arrangements, while the TR is limited just to twin-probe use.

Figure 3.4 Surveying with twin probe resistance measurement. The frame holds a voltage and a current probe, as well as the control box. A cable connects to the remote probes.

The readings obtained will depend very much on the soil under the probes and how wet it is. In winter-wet clays, the resistance may be less than 20 ohms with values doubling as the clay dries out. On sandy soils, the resistance may be hundreds of ohms. What matters is whether a rock (which contains less water than the soil) can be distinguished from the surrounding soil. Of course, that does not tell you whether that rock was laid down naturally or whether it is part of an archaeological structure. That can only be determined by the pattern which a number of rocks form, a subject which we consider later.

You only get a resistance reading when all four probes are in the ground, so when you pull the frame out of the ground, there will be no reading. When you put the frame back into the ground at the next point, current can flow, so you can take another reading. This can be used to trigger the meter, so it only takes a reading after it has seen the 'zero' when the probes are pulled out of the ground. The devices can be set so that they only take a reading once the current has reached a certain level. This reduces the risk of error. However, in very dry conditions, you may have to set the machine to take a reading as soon as any current flows at all and this does increase the risk of false readings.

If you assume that the distance we can see down into the ground is related to the distance between the probes, you can estimate the depth of features detected by using different probe spacings. If you have a series of probes spaced along a line, you can measure resistance between each pair of probes, whether they be next to each other or with a few metres between them. Assume a semi-circle of current between probes, so you see down to half the spacing between the probes (see Figure 3.5). Those pairs of probes next to each other only see objects very near the surface. Those with moderate spacing look from surface to mid depth, and the widest-spaced pair look to maximum depth. You can subtract the near surface measurement from the mid depth measurement to see what is at mid depth, and subtract this from deepest measurement to see what lies below mid depth. You can repeat this along the row of probes to build up a line of information about objects at various depths. It is a little more complicated in reality; you end up with many equations to solve to get the picture, but that is a quite simple job for the right computer software. In effect, you calculate the resistivity of each patch of soil by measuring the resistance of the path through it. You also

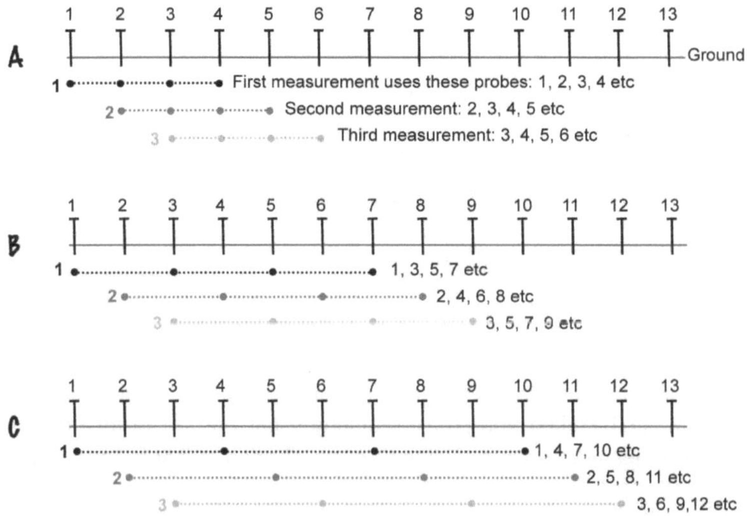

Figure 3.5 Arrangement of probes and connections for depth profiling.

assume that separation is linked directly to depth and this may not be strictly true.

The results are usually shown as a vertical 'pseudo-section' (Figure 3.6, also reproduced as Plate 2 in the colour section), or it would be possible to set up a number of lines, collate the data from each depth and arrange the information in horizontal slices of depth below surface, rather like 'time slices' from ground-penetrating radar (see Section 3.3). This could lead to very big data files but is quite practicable. It may take a long measurement sequence to build up all the data.

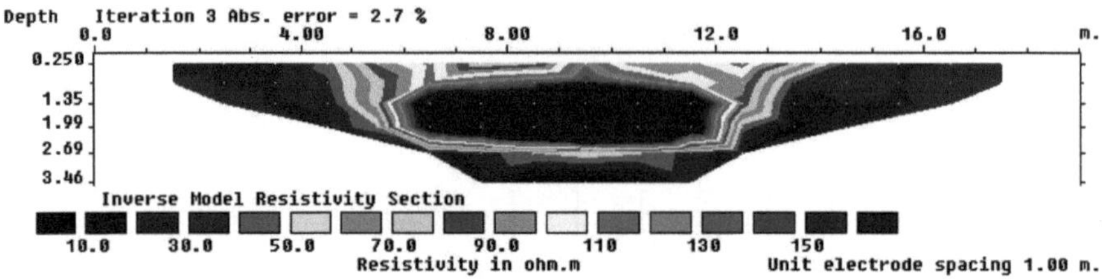

Figure 3.6 Vertical section obtained by depth profiling. The section gets narrower as less readings can be taken along the line at larger spacings. The outer area is low resistance. The inner area shows a section through a buried building. See also Plate 2.

Another electrical resistance property, inductance, is also met in detection methods. We mentioned in Section 2.1 how at one frequency, inductance and capacitance effects cancel each other out, and this is called the resonance frequency. If you choose suitable values of inductance and capacitance, the resonant frequency could be in the audio band, so you would hear a tone if you attached a loudspeaker or headphones. If the inductance were to change, the tone would change, so you can hear that something has been detected.

Any items in the ground, particularly metal, will affect the inductance of a coil held near the ground, either increasing or decreasing it depending on the metal, and this can easily be detected. This is the basis of the metal detector, and some other instruments you may encounter also work this way. Magnetic susceptibility is a measure of the ability to change an inductance, so the presence of materials of differing magnetic susceptibility can also be detected, and these materials include phosphorus compounds deposited by animals and people. Changes in soil resistance can also be detected if coils are set up correctly. Remember from Section 2.3, when discussing light waves, we said that magnetic and electrical effects occurred at right angles to each other as the wave travelled, and this effect has been used for resistance measurement. Such devices are usually called electromagnetic (EM) detectors.

The depth into the ground which a coil instrument of this type can see is about the same as the diameter of the coil. Magnetic susceptibility detectors therefore have very small coils so they only measure deposits on the surface; metal detectors have larger coils which let them see down through the plough soil but no deeper; resistance detecting coils can be large, typically a metre in diameter, to try to see down to that depth.

3.2 Magnetometers

Magnetometers are passive instruments, so you have to be sure that they are detecting archaeology, not you or your surroundings. In Chapter 2, we discussed magnetometers which measure energy and those which measure the field. The latter are the most commonly used, but we will look first at energy devices: the proton and caesium magnetometers.

The proton magnetometer was the first type of magnetometer in regular use in archaeology, but is now somewhat obsolete. Its principal inconvenience was that it was slow. It put the energy into the protons (electrically charged atomic particles) by means of an electrical coil, then switched off the current and used the coil to detect the energy release from the protons over a number of seconds. This meant that covering a large grid was very tedious work. The return was in the form of a sound-like signal, whose frequency increased with increasing strength of signal. Very small signals meant noting a very small change in frequency, but the proton magnetometer could detect down to 0.1 nT in the Earth's field of 50,000 nT.

The proton magnetometer measured the total magnetic field, and gave an absolute value for it. This was a problem where there was strong interference, perhaps from a railway nearby, even a few hundred metres away. One way to overcome this was to have two magnetometers, one well above ground to avoid detecting signals from below the soil, but which sensed the ambient field, and the second for detecting soil signals (and which would also detect the ambient field). If the output of the high detector was subtracted from the lower detector, the ambient signal in each should be cancelled out. However, having a probe well above ground level made for an unwieldy instrument.

This ability to detect the whole magnetic field could be an advantage near the Equator, where the Earth's magnetic field is parallel to the ground, and where the usual downward-pointing fluxgates would not detect any change.

The caesium magnetometer is also an energy device, but it uses light to energise the caesium vapour, and separate types of sensor to detect the released energy change, so it operates much more quickly, typically in 1/100th of a second. It detects the whole magnetic field, so a second sensor may be needed to 'cancel out' any ambient signal. The caesium magnetometer does tend to point in one direction, so it may need to be set up carefully to see the right signals. However, as the frequencies it sees are much higher, smaller changes can be detected, down to 0.01 nT.

Figure 3.7 The 'wobbly cart' belonging to English Heritage. It is designed to carry caesium magnetometers in a non-magnetic, suspended environment.

These types of magnetometer need to be well-cushioned from mechanical shocks and jarring and completely separated from their electronics and batteries. They can be carried in a suitable frame, but are best mounted on a large carriage, which has to be non-magnetic, and pulled along. This makes it unwieldy and expensive kit, more suitable for research at large budget professional and academic levels. Figure 3.7 (see also as Plate 3 in the colour section) shows a picture of a carriage designed to hold four caesium detectors, and Figure 3.8 shows a typical output.

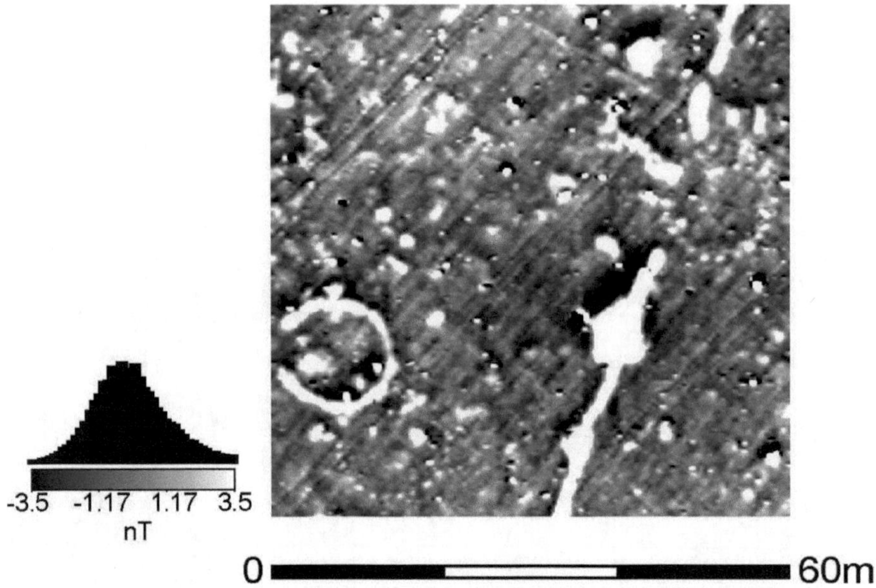

Figure 3.8 Computer output from a caesium magnetometer looks similar to that from a fluxgate device, but it can detect tinier anomalies.

There are more sophisticated types, some of which even use superconductivity at very low temperatures, and these have to be towed on long trailers behind a car. They use a twin sensor pack and each picks up the magnetism of the car equally, so that this effect is cancelled out. This is going to very sophisticated levels, beyond the scope of this book.

Fluxgate magnetometers are much more amenable to general use, and it is most likely that your introduction to magnetometry will be with these. They can be made into compact, readily portable instruments.

Remember that a fluxgate detects the magnetic field, indeed only the portion of the field that is pointing in the same direction as the fluxgate, so which way do you make the fluxgate point? In northern climes such as the UK, the field is pointing steeply downwards towards the ground, and this is the direction where soil anomalies would be detected, so the magnetometers have been developed to look at the vertical portion of the magnetic field. That gives a good response here, but it would be no good near the Equator. You would have to rely on an energy detector there.

If you could carry a fluxgate pointing directly downwards absolutely consistently, you would see the anomalies. However, it is nigh impossible to keep it exactly vertical while walking along. You are bound to swing it a bit and, as it swings, it cuts through the vertical part of the field, and so produces a waving signal. The way to overcome this is to use two fluxgates spaced a little distance apart. Each will respond slightly differently to the anomaly in the earth, but they will both respond identically to the swinging motion. If the two signals are subtracted, only the difference in the anomaly signal remains. This is shown in Figure 3.9. Once you have subtracted two small signals, only a very small amount remains, but these types of magnetometer can detect changes down to 0.05 nT.

These devices essentially look at the change in field with height, that is the gradient of the field, so they are sometimes called 'fluxgate gradiometers'. The use of the second sensor is not quite the same here as in the energy detectors, where the second detector just measures the ambient magnetic field to subtract it. With the fluxgates, both are near enough to the ground to sense the anomalies, so it is measuring the difference between the two detections of the anomaly.

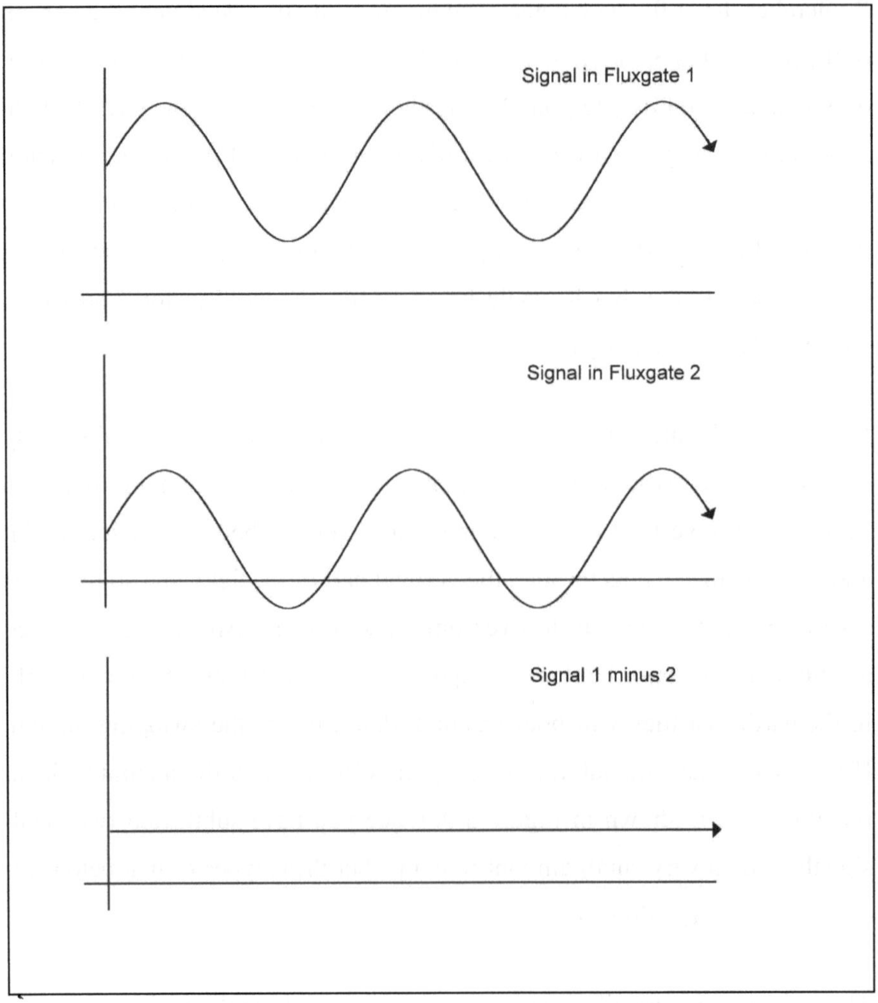

Figure 3.9 The waves show the two signals given by fluxgates at each end of the magnetometer tube. Subtracting these waves gives the plain signal.

This has the disadvantage that the difference between two small signals is itself very small, and this limits the sensitivity of the devices. However, it has the advantage that the detection capability falls away as the cube of the distance: its detection is very close to where it is, and interference cannot be picked up until it is very near by. By cube of distance, I mean distance multiplied by distance and multiplied by distance again. If the distance doubles, the detectable signal goes down by two times two times two, a factor of eight.

There are, of course, magnetic fields set up by electric lines, not just pylons, but also the lower voltage lines to farms, and these quite frequently cross fields. However, these produce fields at very specific frequencies, 50 Hz and multiples of that figure (60 Hz in the United States). These instruments guard against detecting these, as do resistance meters, by switching at a frequency which bears no relation to 50 Hz, so that it always measures a different part of the field, and these different parts cancel out on average.

Figure 3.10 Stripy pattern caused by a metal water pipe.

Figure 3.11 Surveying using a Geoscan FM256 gradiometer. Note the use of a walking string as a guide.

Steel water pipes have their own magnetic field which rotates around the pipe, so sometimes it adds to the field, sometimes it subtracts. This gives rise to a very characteristic striped pattern as shown in Figure 3.10. It may obscure the archaeology, but farmers are often pleased to know exactly where their water pipes are buried!

The practicalities of fluxgate gradiometers are best understood by reference to the current devices, the Geoscan FM256 and the Bartington 601, described in more detail in Appendix A. They look quite different, but they apply the same physics. .

An FM256 is shown in Figure 3.11 (see also Plate 5 in the colour section). It comprises a vertical aluminium tube just over half a metre long and 50 by 50 mm section, connected by a handle to an electronics head. The head and its display are shown in Appendix A3. The tube holds two fluxgates, top and bottom, half a metre apart. The aluminium tube is rigid but lightweight. There is a problem with aluminium in that it expands as it warms up, and this could cause a change in shape, slight but significant. This means that you have to leave the mag to settle to ambient temperature before using it (about 15 minutes), and you may have to recalibrate the instrument regularly when the sun is strong or there are big temperature changes.

As the two fluxgates have to be aligned exactly, two very fine screws allow them to be moved slightly to get the alignment right. One adjusts 'north–south', the other adjusts 'east–west'. When it is set up correctly, the display shows the same reading whichever way the magnetometer is pointed. A control ensures that the two fluxgates give the same output, and this has to be tested by turning the device upside-down and checking that the readings are still the same. You do these operations manually with an FM256; they are done automatically on the Bartington. Figure 3.12 (a–d) shows the different orientations needed to hold and check the FM256 in order to calibrate it.

The head provides a display for showing the signal level relative to an arbitrary zero. You actually set the zero during calibration. It does not show

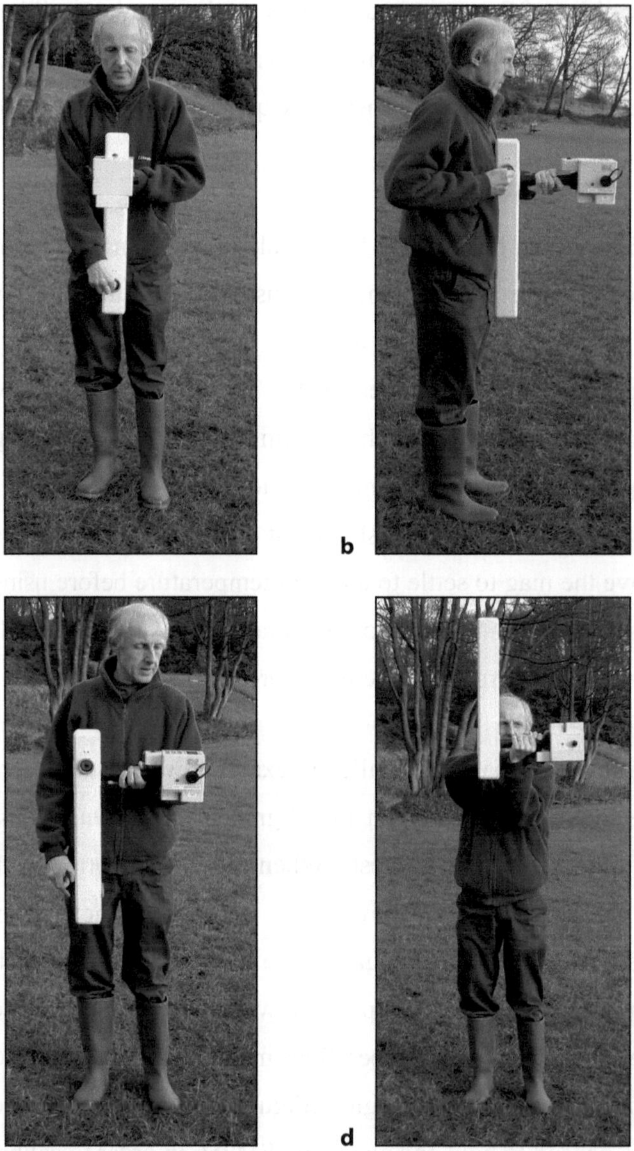

a b c d

Figure 3.12 Calibrating the FM256. This involves holding the head first north (a), then south and adjusting to the average of the two readings; then measuring east (b) and then west and averaging, and then measuring upright (c) and upside down (d) and averaging.

the whole magnetic field, so the reading is not 50,000 nT, but typically –1 nT or + 5 nT. The display also shows grid details for data logging control (see Appendix A3). The touch buttons on the head allow for setting up grid details, sensitivity, logging or clearing the memory and downloading as well as setting zero. You can either set the device to log automatically so many times a second, in which case you have to walk over the grid at a very even pace, or you can use a manual trigger to take a reading as you pass a mark on the ground. (Gridding is considered in detail in Section 5.3.) You carry the instrument by its handle. It does not matter which hand or which way round, but you must be absolutely consistent, for instance always keeping the head pointing west, whichever direction you walk in. When you turn round at the end of a line, you should turn yourself round without turning the device. Transfer it from one hand to the other.

The Bartington 601/2 is shown in Figure 3.13 (also reproduced as Plate 4 in the colour section). It actually comprises two fluxgate gradiometers one metre apart. Each of the gradiometer tubes contains a pair of fluxgates one metre apart vertically. Increasing the vertical separation means a bigger signal difference for the same gradient, so greater sensitivity. Once you have such an unwieldy device to carry, it is easier to support it in a harness than to carry it by hand; indeed it is easier to carry two devices, one on either side for balance. Having two gradiometers a metre apart allows you to do two lines at once – half the walking for the same area covered. There is, however, a single tube device for the less wealthy and more leisurely.

Once you have a harness like a pair of handlebars, you can easily mount the display and control console centrally on it. It has similar functions to that of the Geoscan magnetometer, but controlled by menu rather than by touch buttons. It also has to be calibrated in a way physically similar to the Geoscan FM256, but this is done automatically as the display gives you instructions on where to point it. You also have to turn the whole device round with yourself at the end of each line. The Bartington 601/2 is described in more detail in Appendix A4.

Figure 3.13 Surveying using a Bartington 601/2 dual gradiometer. Each tube is an individual gradiometer, so this device can survey two lines at once.

3.3 Ground radar

Microwaves typically have wavelengths of a millimetre to a metre, which are normally associated with radar. A short pulse of radar signal may therefore be less than a metre long, so that we can resolve objects to this detail. If we can send a radar pulse down into the ground we can await a reflection, and if we time how long it takes for the reflection to arrive, we can calculate the distance the pulse has travelled there and back, so we know the depth of that object is half that distance, as shown in Figure 3.14. This is very similar to echo

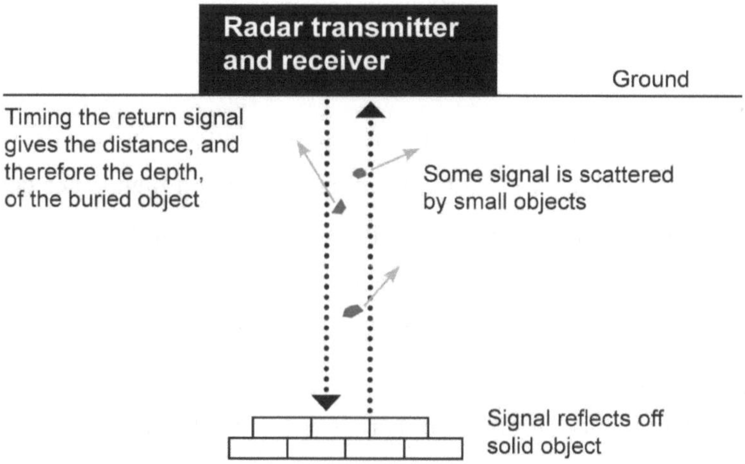

Figure 3.14 Radar sends a pulse into the ground and detects what is reflected off an object. The journey time of the pulse is the time down plus the time back up.

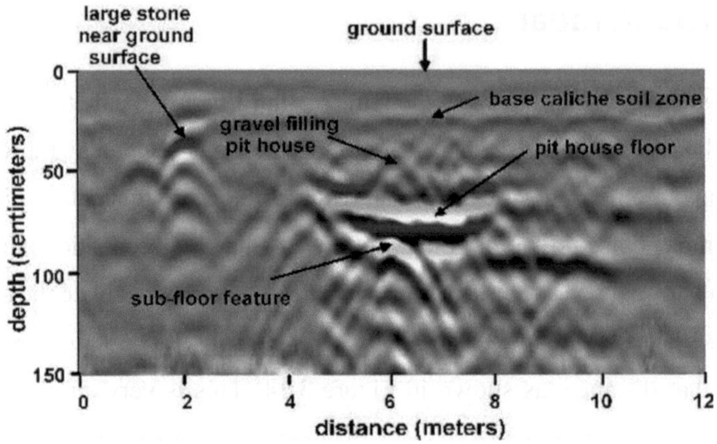

Figure 3.15 A typical radar screen, but which has been annotated by the user. Items appear as an upside down 'U', as the radar sees them first at a slant, and then closest as it passes directly overhead.

sounding used by boats, except that boats use sound waves through the water. The ground into which we are transmitting will have a high water content which will slow the wave, give too shallow an estimate of depth, and it will also absorb the microwaves so they cannot travel far. However, the wave speed can be calculated for different types of ground and the depth estimate corrected. The greater the water content, the greater will be the absorption, so radar will work better on dry sand than on wet clay. The absorption effects can be reduced by using longer wavelength radar, but that will mean a longer pulse, so detail will be lost. The speed of light in the ground is reduced typically to about one quarter of its value in air, so the wavelength is reduced by a similar amount. For example, a wavelength of 0.8 m in air is reduced to 0.2 m in the ground.

The object we are seeking needs to look very different to microwaves from the ground it travels through, for instance a layer of rock, so that the waves are reflected off it. We must also remember, though, that some waves will be scattered, perhaps off soil particles. These will get back to the detector and produce false but faint signals to confuse the picture.

We also have to remember that the pulse of radar is not infinitely thin. Also, it spreads out a little to either side, so that an object will be seen when it is still ahead of the radar, before it passes underneath, and then it will be seen a little way behind the radar. When it is directly under the radar, it will be at its closest, and will be observed at its true depth. In front and behind, it will appear at an angle, and so apparently further away. This gives rise to an 'upside-down letter U' shape at the detector, as shown in Figure 3.15 which illustrates a typical chart of signals from a single sweep of radar, with the main features annotated. This effect can be reduced by using larger transmitters and detectors, but if these are too large, they will upset the layout of the device, and give other errors. If we can recognise what is happening, we can process the returns by computer to remove the slopes and leave the true points.

Colour may be used to enhance the picture by showing different strengths of signals, but there is no colour in the original. Note that this is just a narrow sweep, perhaps a metre wide, and ideally we need a number of sweeps side by side covering exactly the same amount of ground, so that we can see the signals from a much wider area. By the time we have achieved this, we will have accumulated a huge quantity of data. Memory capacity may limit the area covered, and the sheer complexity of the picture in three dimensions may be beyond our understanding.

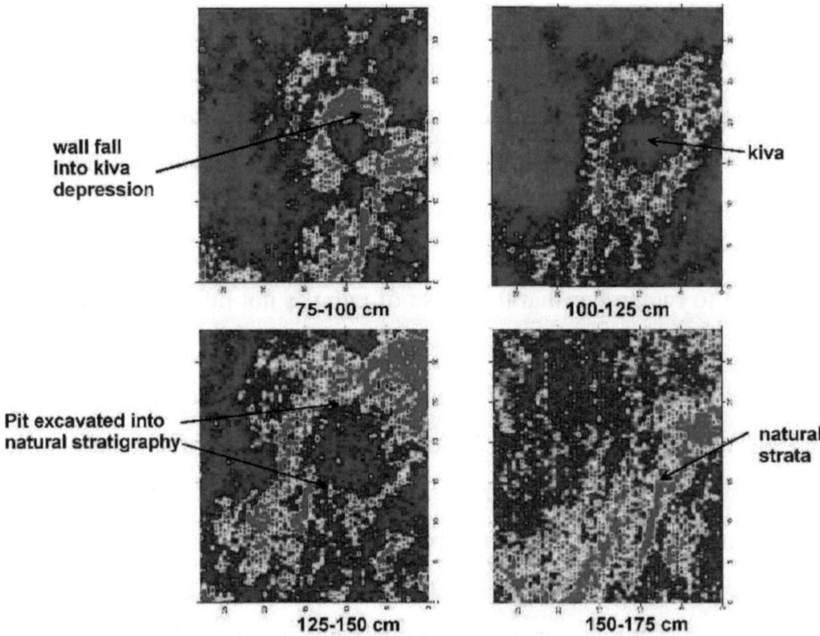

Figure 3.16 A set of time slices obtained by radar of an Indian stone dwelling, or 'kiva'. Each slice represents a different depth under the surface, so you can see at what depth the archaeology is concentrated. Here it is mainly in slices 100–125 and 125–150 cm. See also Plate 6.

The easiest way would be to look at the length and width of the area covered by the sweeps and plot all points which returned a signal after the same time interval. This will give us an area at a single depth under the surface. It is often known as a 'time slice' as it is a map of all points with the same radar return time, but note that it is not a slice of time in the archaeological sense. A later archaeological feature could be cut down into earlier layers, and the wall of an earlier building could stick up into more recent contexts. A time slice is illustrated in Figure 3.16 (see it also as Plate 6 in the colour section). Interpreting a series of time slices requires care.

Figure 3.17 A typical radar set ready for use. The controls are on the buggy, with the antenna towed behind, and the wheel behind that measures distance.

A typical type of ground radar is shown in Figure 3.17 (see it also as Plate 7 in the colour section). The radar is controlled by the console and screen high on the trolley. The antenna is dragged along behind the trolley, with connections to the console and the battery at the bottom of the trolley. The wheel behind the antenna is of known size, so that distance moved by the radar can be measured as it turns.

Probes can be made which work at different frequencies and wavelengths. The most common probe operates at a frequency of 400 megahertz (MHz);

that is a wavelength of 0.75 m in air, but significantly lower through ground. High frequency, high definition radars may go up to 1500 MHz, shorter wavelength. Low frequency, high penetration radars may go down to 100 MHz, longer wavelength. Lower frequencies will need larger probes if they are to give clear pictures, as the width of the radar beam depends on the ratio of the probe size to the wavelength. Radar for use in air usually works at higher frequencies (shorter wavelengths), but these do not have to contend with ground water.

The smoother and dryer the material to transmit through, the greater the capability of radar, so a thick stone floor may be easier to search through than a grass field.

Given the huge amounts of data generated, it is also best to use radar in the field for probing a localised site first detected by other geophysics techniques, such as magnetometry, which are better suited to large area coverage.

3.4 GPS, laser theodolites and total stations

GPS, laser theodolites and total stations all rely on light waves, but not visible light. GPS stands for Global Positioning System, which is an American system for locating yourself at any point in the world. It was originally created for defence purposes, but is now widely used in many commercial fields. Russia and the European Union are also producing their equivalent systems, Geomass and Galileo. GPS relies on a number of satellites. These

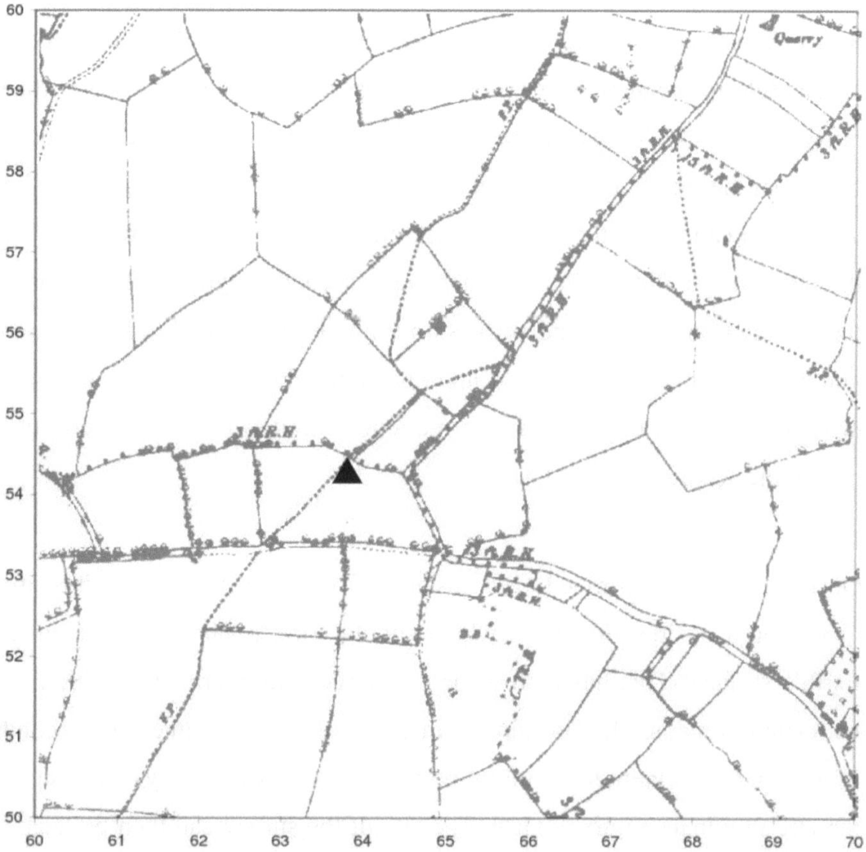

Figure 3.18 Reading coordinates. First read the 'easting' (the bottom set of numbers), then read the 'northing' (the side set of numbers). Note this is a fictitious example applied to a nineteenth-century map.

orbit the Earth, but all at different angles to the Equator, so that they fly over different parts of the Earth. If you have a detector which can receive signals (electromagnetic waves) from at least four of them, and one of them provides a clock, you can work out where you are relative to the satellites by measuring how long each signal takes to reach you. The more satellites which you can track, the better the fix on your position. The waves travel through space at 300 million kilometres per second, so you need to measure time to the nearest one hundredth of a millionth of a second to know where you are to even a few metres. A more accurate clock might help, but other problems, such as the slight speed reduction through the atmosphere, then reduce the accuracy.

Once you have determined your distance from each of the satellites, that gives you a position in space, which should correspond to a point on the Earth's surface! This has then to be converted into coordinates which can be related to maps. The coordinates might be degrees, minutes and seconds of longitude and latitude, or might be a country's local grid system, such as the UK's Ordnance Survey Grid. The site of the survey can then be located on maps. Figure 3.18 shows how a grid reference is given. Height above sea level also comes out of the equations.

Figure 3.18 is a fictitious example, showing a nineteenth-century map overlaid with coordinates. Always read the easting (the horizontal scale) first. The arrow is between 63 and 64. If you measured it more exactly, you would say 638, and even more exactly, 6379. Now read the vertical scale. The arrow tip is between 54 and 55. Reading it more exactly gives 543, and even more exactly 5432. You would give a six-figure grid reference as 638543, or an eight-figure grid reference as 63795432. There will be a prefix, such as HF or ST, to show which part of the UK you are in.

Small, low cost, hand-held GPS meters and GPS on mobile phones are useful in archaeology for the purpose of locating surveys on to maps, but they are not sufficient for laying out a grid for a survey. The meters also provide a value of height above sea level. Again, this can be useful when relating a survey to a map, but calculation of height is more tenuous and any reading is best checked against a contour map to ensure its accuracy.

Much more accurate measurements can be made, precise down to a few millimetres, but these require a Differential GPS system, a much more sophisticated and expensive arrangement which compares the GPS location of your detector to that of an accurately located ground station. Calculating the position of each of the two sites will be subject to the same errors. If the positions are 'subtracted' one from another, the errors are also subtracted, leaving a very accurate measurement of position. The ground station could be up to 50 km away and independent, but often a complete system can be set up with fixed station and roving unit for you to measure in your points. This is one form of a Total Station. Such a ground station and roving unit are shown in Figure 3. 19.

The equipment described above makes you reliant on satellites. There are times when you may not require or want such detail, but you may want to know the boundaries of your survey or the boundaries of the field the survey is in. You may also want to draw a contour map to fit to your survey. It is all very well to discover a ditch, but it also helps to know which way water would flow in it and how steep the ground is.

a

Figure 3.19 A GPS total station: (a) the roving unit, (b) the ground station.

b

There have been suggestions of adding GPS locators on res and mag instruments so that they can log position at the same time as they log data. This could then all be downloaded to the computer to produce the mapped-out data. This would require high accuracy GPS which would add cost and bulk beyond simple use. It also has the disadvantage that it only tells you where you are when you log the data, not where to go next to log the next point, so without good discipline you could end up with a plot very dense in some places, very sparse in others. This is where the grid set up described in Chapter 5 helps ensure a good even collection of data.

A different form of total station could be used, perhaps at a less sophisticated level, and an electronic distance meter (EDM) can also be utilised. An EDM is a theodolite (a traditional high accuracy survey instrument) fitted with a laser. These still use electromagnetic waves, but in a different manner. The main part of the instrument is a laser which fires a wave pulse and counts the time until the pulse returns after being reflected. The laser is usually a low power, infra-red device, but which allows ranges up to a few thousand metres. The time of travel can be converted to a (two-way) distance, knowing the speed of light. If the distance to the reflector is then known, the machine can calculate its grid coordinates if it knows its own position on the grid and the angle it is pointing, relative to grid north. If it also knows its angle of tilt, and its height above sea level, it can calculate the height of the reflector. This is shown in Figure 3.20. If the EDM can rotate and tilt automatically to find the target, and calculate its position while the operator carries the target around, this is also known as a total station.

Some systems are advertised as 'reflector-less', meaning that the detector can pick up enough light returned after bouncing off any surface. However,

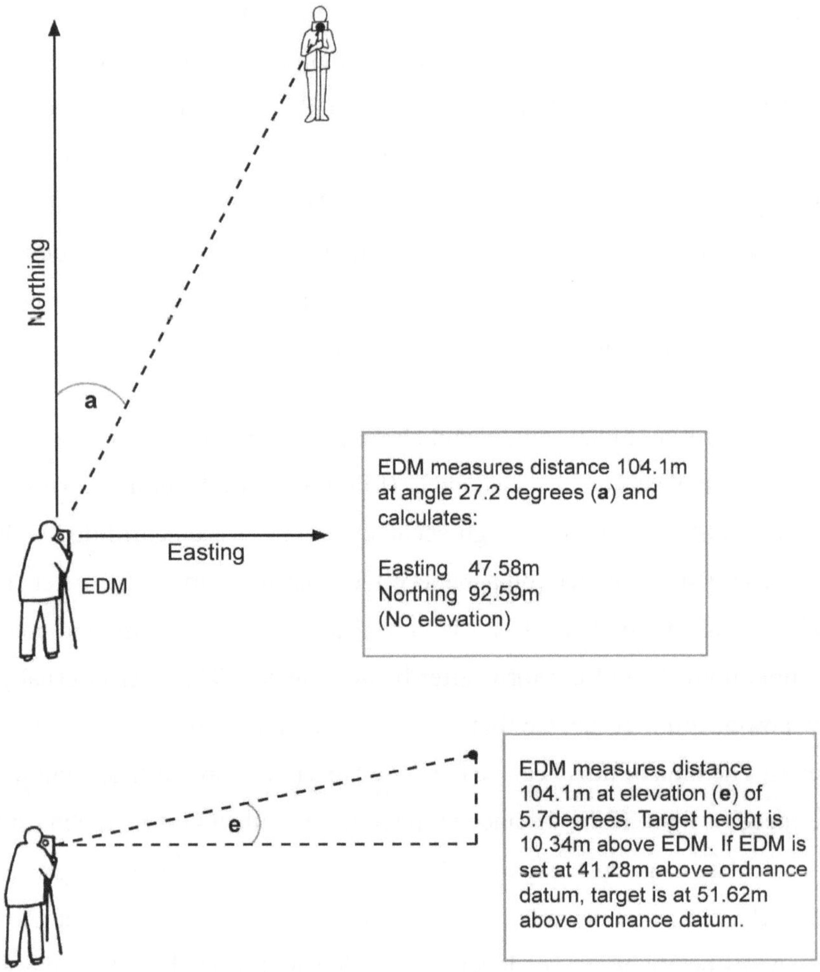

Figure 3.20 The electronic theodolite measures distances and angles, and can convert them to eastings, northings and height.

these are not best suited to archaeology. It is best to use a reflector of the sort shown in Figure 3.21. It can then be placed accurately to measure the exact spot you want measured, and there is no risk of a stray reflection giving the wrong position.

With a sophisticated total station, you walk around the points you want to measure, and the device tracks round to follow you, taking and logging a

Figure 3.21 The target is on the pole held steady at the point to be measured.

Figure 3.22 Electronic theodolite. The operator uses a telescope to point at a target. The laser above the telescope fires a beam to the target and detects the return. The electronics under the telescope calculates distance, angle, vertical angle and position of the target and logs the data.

measurement when you command it. On older, simpler systems, EDMs, such as shown in Figure 3.22, you walk round with the reflector, while a colleague makes the measurements with the device (Figure 3.21) and either logs them or writes them down. If a colleague is operating the EDM, it is essential that you have walkie-talkies or a pre-agreed set of arm signals that are clear and distinct over several hundred metres. It is also essential that the operators, particularly the reflector carrier, are highly visible to each other. On a fine day, you may see someone wearing a dark jacket wave at 50 m, but

Figure 3.23 On a clear day, you can see the target as on the right, but on dark winter days and at large distances, you may only see the handler's reflective jacket (left).

on a dark, damp winter afternoon, they are almost invisible at that distance and it is impossible to locate the target through the telescope. Hi-visibility jackets and flags are essential for good visibility and signalling. This is illustrated in Figure 3.23. Compare the two parts and note how little other than the jacket can be seen in poor visibility at long distances.

EDMs and total stations can operate to several hundred metres, but they may be limited at very short ranges. If you are only measuring to the nearest

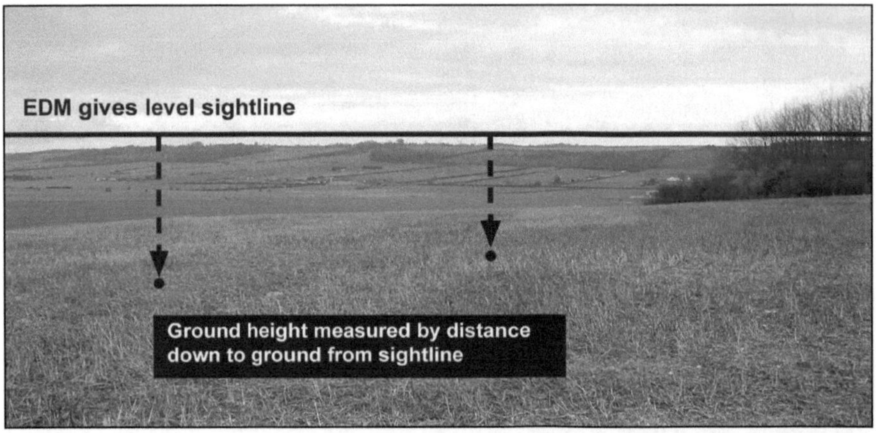

a

b

Figure 3.24 The electronic theodolite must be set up exactly level so it sweeps a level line (a). If it is not level (b), the line will slant, and give wrong heights.

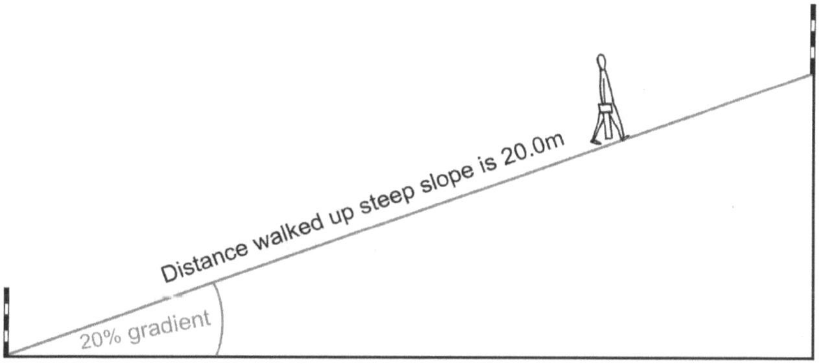

Figure 3.25 Steep slopes can cause surveying problems. The distance you actually walk up the slope is more than the 'level' distance. Which do you use in the survey?

hundredth of a millionth of a second, that is a distance of over three metres there and back. There may a bit of 'dead time' when the device cannot receive after transmitting, so such equipment may only be able to measure distances from five metres upwards.

When setting up a total station or EDM, you need to feed grid coordinates to the static part, whether they be national or just for your own grid. You may call the point where you set it up (0 east, 0 north) or, better, (1000 east, 1000 north). The latter will usually avoid negative numbers in the coordinates, which can lead to confusion if you are not meticulous with 'minus' signs. The device also has to be levelled very precisely or its calculations of height and grid coordinates will be wrong, as the arc swept by the instrument will not be flat, but will point too high to one side, too low to the other side. This is shown in Figure 3.24.

EDMs and total stations also convert readings to be level. That is, they ignore any slope and just calculate horizontal distance. If you measured that distance with a tape measure, you may get a slightly different value, as the tape follows the slope of the ground. Differences are usually small, unless the ground is steep. This is shown in Figure 3.25.

We have now looked at the application of science to the operation of geophysics equipment. In the next chapter, we will look at downloading data from the instruments to computer.

4

Processing the Data

4.1 General

Once the data have been collected in the field, they need to be presented in a way which lets you understand what is under the ground. This is usually best visualised by producing a plot akin to a map, so that the detected features appear flat on a screen which has a known scale and north direction. Therefore, if a building appears a few centimetres above the centre of the plot, slightly to the right-hand side and the scale is "1cm equals 20 metres", this would show that the building was about 20m east, 30m north, of the centre point of the field. It can then be located for any excavation work. This is shown in Figure 4.1, where a building appears faintly.

There are a number of software packages which can provide such a picture, even on a laptop computer in the middle of a field. If you were so minded, you could even write your own package, providing it does the same tasks.

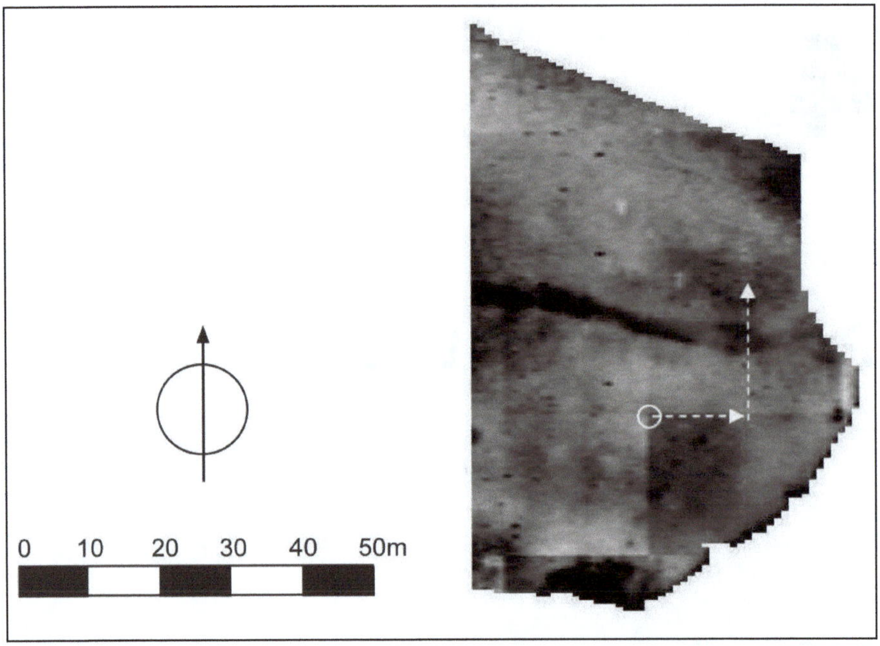

Figure 4.1 Locating a feature (a building) from the centre of a grid. Shown on a geophysics printout.

Some of the most common software packages are described in more detail in Appendix B.

Such packages perform three distinct tasks. The first is to transfer the data from the instrument to the computer. The second is to rearrange the long stream of data from the instrument into a pattern which mimics the pattern you made when walking up and down the field. The third is to make the data intelligible. In some cases, different software may be used for task one than for tasks two and three.

Each make and type of instrument will have its own way of storing the data, of talking to the computer, and its own connector type. Before starting field work, it is important to make sure that the computer has the correct types of connector and the software can speak to the particular instrument. The software also has to be compatible with its host computer. Software is often supplied with some sort of key, either a sequence of numbers and letters entered, or a 'dongle' which plugs into the back of the computer. This means it cannot be distributed to many computers indiscriminately as it only works when that key is present.

The first task is apparently the simplest, but can cause the most problems. Each software package was designed to speak to certain instruments which were in regular use when it was written. A package will not necessarily speak to a more modern instrument, or even to obsolete instruments which are still in service. Your choice of instruments may make it necessary to have different software packages to speak to each. This is where it can help to separate the communication from the processing. However, the data sent to the computer has to be compatible with the input requirements of the processing software, and this can cause further problems of data conversion.

We will look first at communicating between instrument and computer, then at laying out the computer screen to represent how you have walked the field, and then at processing the data and presenting it. We will concentrate on resistance and magnetometer measurement. The concepts will be similar with other instruments, but may be very much more complex – for instance in displaying ground radar data, which has depth information as well as a flat pattern. In this case, the amount of data to be handled may become very large, so that understanding how you are sorting it becomes very important.

4.2 Talking to the instruments

While you have been walking up and down the field, the instrument has been logging the data. That data then needs to be transferred from the instrument into the computer and sorted to represent the pattern you walked. You will normally have divided the field into individual grids and set the instrument to accept the number of points in your grid before starting another grid. For instance, if your grid was divided into 20m squares and you took a reading every metre along lines one metre apart, each grid would contain 20 times 20, that is 400 data points. The instrument may record the position in the grid in terms of line number and point on that line as well as the reading or it may just record a long series of readings. You will almost certainly have walked up one line and down the next, in a 'zig-zag' fashion. You will need to know if the instrument has followed your footsteps in recording the data or whether it has counted up as you walked up a line and then counted down as you walked back. If it has done that, and you do not realise it during download, your picture will be nonsense, as alternate lines will be the wrong way round. If you have set the instrument to record data in a particular sequence but not followed that sequence while walking your results will also be nonsense, but that will be your fault.

You also need to know how to cope with partial grids. There may be a portion of a grid left before the field edge after you have walked the last complete grid of a line. If you want complete coverage of the field, you will need to walk as much of the partial grid as you can, and then put in null values for the points beyond the field edge which are in the grid. You may also need to put in dummy readings in a magnetometer survey as you approach iron objects, such as wire fences or derelict farm machinery. You need to know what value the instrument puts in for a dummy reading, so that you can ensure that the software recognises it.

Downloading – that is, getting the instrument to talk to the computer – normally consists of connecting the two with a special cable, and then pressing the button on the computer and a button on the instrument, but even this can have a number of problems. The special cable will have the right connectors for the download, but so does the computer. Many instruments use a connector known as RS232D, but the equivalent connector is not fitted on many new computers. You may need an old, second-hand computer to talk to the instrument. If the software is protected by a dongle, you may also need a parallel printer connector not found on many new computers. On the other hand, if you have a new instrument with USB2 connections, an old computer may not be equipped for talking to that instrument. Computer connectors are shown in Figure 4.2 (see also Plate 8 in the colour section).

a **b**

Figure 4.2 Connectors on a computer (a) connected, (b) disconnected. From the left, USB connector and memory stick, then (other side of mouse connection) RS232D and download cable, then parallel connector and dongle.

You will need to know something of the expectations of both software and instrument. Will the data come across in the format that the software is expecting and at the expected speed? If instrument and computer have been used together before, this is not normally a problem, but it is a useful check if you have to sort out why the two will not talk to each other.

You will also need to know what the software expects the data to look like. For instance, does the software just look for a gap in the number sequence while the instrument puts in a comma? If the software has built-in facilities for recognising the instrument you are using, there will not normally be a problem, but if you introduce a new instrument, or have to download data and then import it to the software, there can be problems of software not recognising the data as the format is wrong.

You then need to know the correct button press sequence. For most instruments, you press a button on the computer and that tells you to press a particular button on the instruments. However, in some cases, such as the TR/CIA resistance meter, you must press the button on the instrument first, or it will not be ready to talk to the computer and nothing will happen. This will be considered in Appendix A2 where it is described in more detail.

Where are you going to store the data before giving it to the processing software? If software packages can talk directly to the instrument, you can select your project first and ensure that the data are directed to it. Otherwise, you may have to create a folder to contain your data, which you can access from within the computer. The data will usually comprise a number of files, each representing one grid. If you are using more than one instrument, you need either to store the data from each in different folders or give each set a prefix, for instance 'r' for res, 'm' for mag.

You will also need to renumber the grid files after day 1. If, for instance, your folder has 12 grids from a previous day's work on that field, you will need to continue the sequence by telling the download software that today's grids start at number 13. What the instrument then considers to be 'grid 1' will be stored on the computer as 'grid 13', the instrument's 'grid 2' becomes the computer's 'grid 14' and so on.

Sometime after you have completed the download successfully, you will need to delete the grids from the instrument's memory, but it is a good idea not to do so until you have processed the data and ensured that it is safely copied across and stored. It is also wise to back up the data to a second device, for instance a memory stick, before you delete the data from the instrument.

4.3 Arranging the data pattern

At this point, we assume the data are all correctly transferred to the computer and in a folder where the processing software can use them. The folder may be within the software ready for immediate use, or the data may have to be imported into the software. You then need to tell the software about the contents of each grid and how to position them on a map so that they represent the pattern you walked on the ground. This differs so much from one software package to another that only the general principles can be considered here.

Sometime during the data transfer, you will have told the computer:

- how many data points are in each grid;
- how they are arranged in terms of number and spacing of lines;
- how they are arranged in terms of number of points or spacing along each line; and
- whether or not to lay the data out in 'zig-zag' pattern.

You also have to know the starting point (e.g. south-west corner) and direction (e.g. northwards) of the first line. Each grid then needs to be placed on a map. In some software, this may mean giving a whole set of instructions to place grids so that the start is, for instance, at the bottom left corner. For another, it may involve filling numbers on a spreadsheet, relative to the grid start point. On some software, you may be able to place the grids in any order in a pictorial manner, even if the start points are mixed and directions vary. The main thing is that the pattern must accurately represent how you covered the ground, both in grid content and grid number. Then the data are ready to process.

4.4 Processing the data

Ideally, the picture presented will show a map of the features below the ground you have walked. However, the first view of the data presented may not look anything like as good and may even leave you dismayed. The purpose of the processing is to remove the imperfections in the collection of data to provide a clear picture. You can even do your own processing, perhaps using a spreadsheet, providing you can handle the very big numbers of data points and arrange them into a pattern like the pattern you walked over the ground. You then need to know what causes the imperfections before you can find a way of eliminating them.

This may sound a little like cheating, but there are a number of effects, particularly in mag, which can degrade data collection, especially if the operator is not experienced. Your first sight of the downloaded data may look as awful as Figure 4.3, but if there is real archaeological information there, you have to remove the clutter so that you can see it.

The picture shown in Figure 4.3 is mag data, which is the most prone to imperfections. Two forms of imperfection are obvious. Some grids are stripy, and on others, there is shading from one side of the grid to the other. The latter is generally caused by a drift in calibration so is an instrument fault, but it still needs sorting in the software. It is too late to go back and repeat the grid, although that may occasionally be necessary if the drift is severe, more like instrument failure than imperfection. Sometimes, such shading is caused by the operator getting within a few metres of an iron fence or a car. It can still help if the software can remove the effect.

The striping effect is very much an operator problem, especially among novices. Holding and handling the mag will be discussed in more detail in Chapter 5, but it is usually the case that the operator has not carried the mag in a steady fashion after turning round. It may be that they have not removed all metal and are very slightly magnetic on one side, they may be carrying

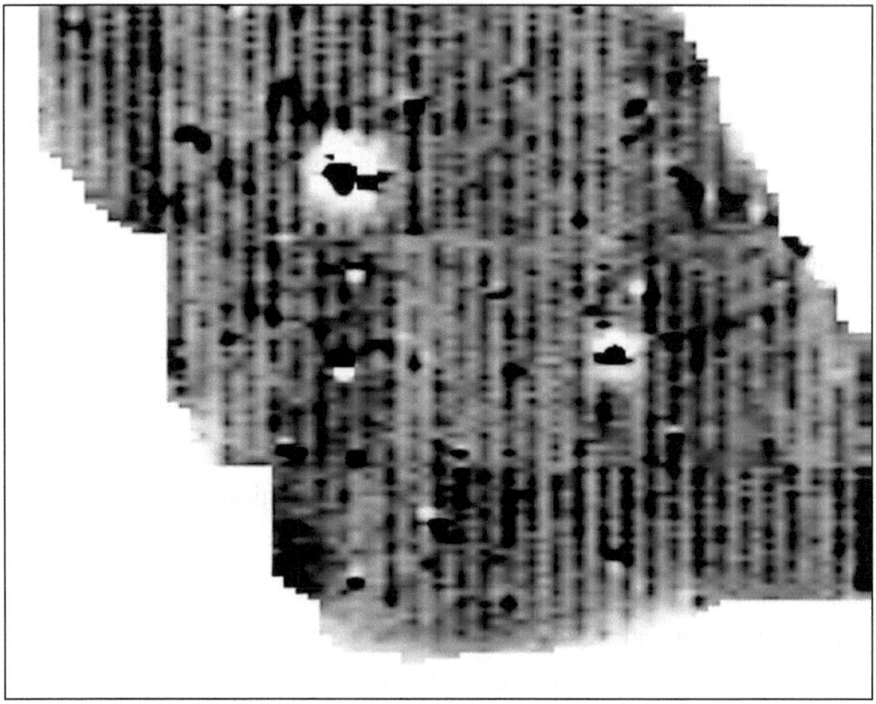

Figure 4.3 Stripes in the magnetometer data hide the faint features.

the instrument at quite a tilt, or they may have let the mag swing round with them when they have turned at the end of a line.

Different software packages will have different ways of improving the picture. Some will use advanced mathematics to compare ends of the grid or neighbouring lines on the grid and then smooth the results to give as near uniform lines as they can. Other software may reduce the average value of each line to zero, so that features can be identified as the non-zero readings.

Providing the real signals are quite strong compared to the imperfections, a good map can then be obtained, but if the signals are weak or striping very strong, it may not be possible to clean up the picture very well.

Other imperfections can occur which may need to be ironed out to produce a clear picture. Sometimes, one boot can be slightly magnetic, so that the readings change according to whether the left or right foot is forward. Another bad effect can happen if the operator does not start and finish at the right point, but strays by a metre or so at one end. Then a feature will appear to zig-zag across the grid rather than form a straight line because the lines did not start and end where the software expected them to. Software packages will usually have means to adjust the data display for these, but they need to be used with caution if you are not to produce a false picture.

You may also need to cut out high spikes, perhaps because you have inadvertently walked over a buried iron plough blade which produced a very strong magnetic effect. The software will normally have a means of removing these, either by chopping off the very high readings or by removing the data points and putting in blank data. The black 'blobs' surrounded by white in Figure 4.3 look like effects of buried iron.

When you have finished the processing, the data may now look a lot different (Figure 4.4), and you can see if there are any archaeological features. Note the large 'iron' signals have not been removed here.

Although res is less prone to handling and calibration errors, it can still need processing to get the best picture. A problem can occur when moving the remote probes. If they are set at a different spacing, or set over some stones, the measured values of resistance will change, so that, for instance,

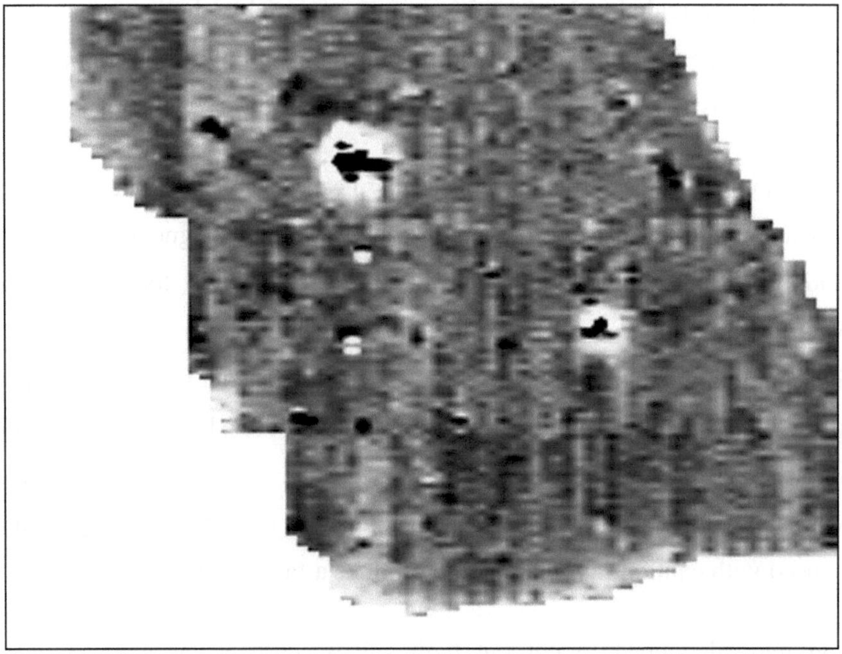

Figure 4.4 Once the stripes have been evened out, a faint circle appears just left of centre, with lines from it down and to the right.

an ordinary reading may be 20 ohms in one grid, and 30 ohms in the next grid. That next grid will appear darker on the plot, but it is not indicating any archaeology. You need to have a means of adjusting the levels of each square so that the features match, rather than the absolute values of the readings.

The readings can change across a field by a large amount anyway just because of the nature of the ground, not because of any archaeology. The soil may be a lot deeper at the bottom of a slope than it is at the top, so you may get much higher readings at the top of a field just because the naturally occurring bedrock is near the surface. You may have to reduce these levels compared to those at the bottom of the field so that you can see any archaeology, even if this means having dark bands or areas on the map which you cannot match across all grids.

You will probably still need a means of removing big spikes. These can easily happen in res if a probe hits an isolated stone or if a reading triggers before the probes have made good electrical contact. Thcsc arc best taken out and remeasured as you walk along with the res, but you may not notice all of them in time. Using the res will be considered in more detail in Chapter 5.

Figure 4.5 shows a map from res after processing. Note that the shading does not necessarily match from one grid to the next, but that is because there are

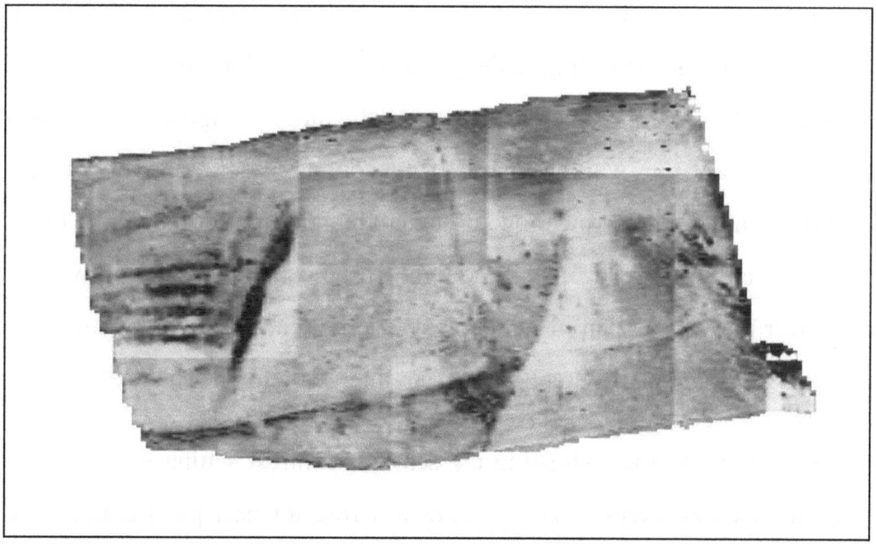

Figure 4.5 Resistance plot, showing a field with stone drains (left), and a wall which goes to a small house (bottom centre).

not enough shades of grey to keep to one scale and show the archaeological features as well as the natural features. Some processing software packages provide the ability to link grids so that there is continuity of shading. That may look better, but it masks the archaeology as it adjusts scales and levels to suit the whole picture, not just the interesting part.

4.5 Display

After processing, you get your first clear view of the archaeology (or its absence). Then you have to know what to make of it.

Firstly, you may want to alter the scales of signal strength to bring out particular features. According to the package, you may have a chance to alter the scale or the contrast. The latter refers to the range of signal values which covers the span from white to black. If the contrast is set high, you will get a lot of detail over a narrow range of signal, but all high levels will be black with no detail, all low levels will be light. If the contrast is set too low, you can cover the whole span of values, but lose intricate detail as two distinct readings representing different features are merged into one shade of grey.

It is usually best to try adjusting the scale or contrast setting to see which gives the clearest picture. There is no firm rule for best picture and there may be no setting which is best for the whole survey area, so find the best compromise.

The scales used in the picture may be 'linear' or 'logarithmic'. The latter usually gives best detail, but it is worth knowing the difference. A linear scale just adds a fixed amount for each step in the scale, so that it might go: 10, 12, 14, 16, 18, 20… for equal steps.

The logarithmic scale multiplies each step by the same amount, so the steps would be: 10, 10*1.2 = 12, 12*1.2 = 14.4, 14.4*1.2 = 17.3, 17.3*1.2 = 20.8, 20.8*1.2 = 25… The logarithmic scale has covered a bigger range of readings, but the steps at high readings have become larger, so detail might be lost. It distinguishes between 'big readings' and 'small readings' but is less good between. Adjusting the contrast alters the multiplier. With lower contrast, using a multiplier of 1.4, the steps would be: 10, 14, 19.6, 27.4, 38.4, 53.8… while the linear scale at lower contrast would give: 10, 14, 18, 22, 26, 30…

Unless you know values of readings of particular features, it is better to experiment with settings rather than try to predict the best setting.

The scales shown would apply to res, but in the case of mag, the scale would more likely be set to give steps between –10 and +10, with zero being a central value.

I have assumed above that the picture of the archaeology has been displayed with black representing high readings, white representing low, with various shades of grey in between. This is the most common display, but not the only one. Sometimes, the scale is reversed, so that black represents low readings. This can be helpful if ditches show in res, as they then appear as dark lines. Alternatively, colour displays can be used. For instance, the scale may go from red to black or even from pale blue to red, through a whole range of

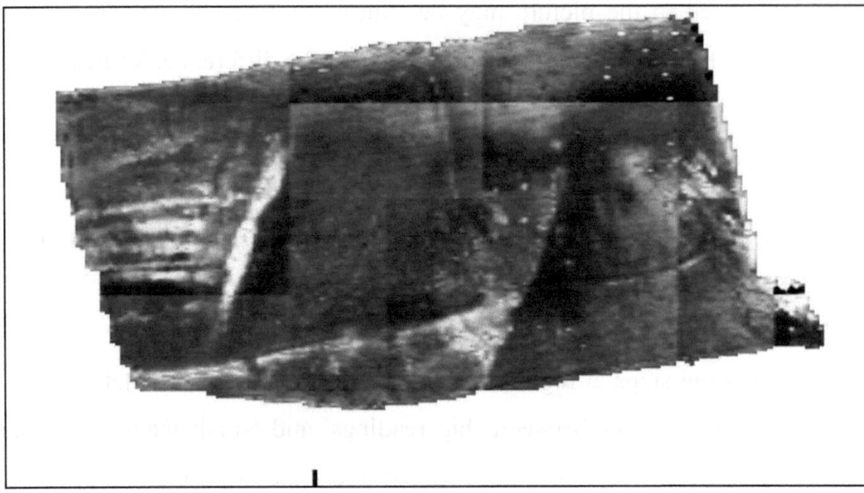

Figure 4.5 The same plot as Figure 4.4, but with the scale reversed so that white represents high resistance.

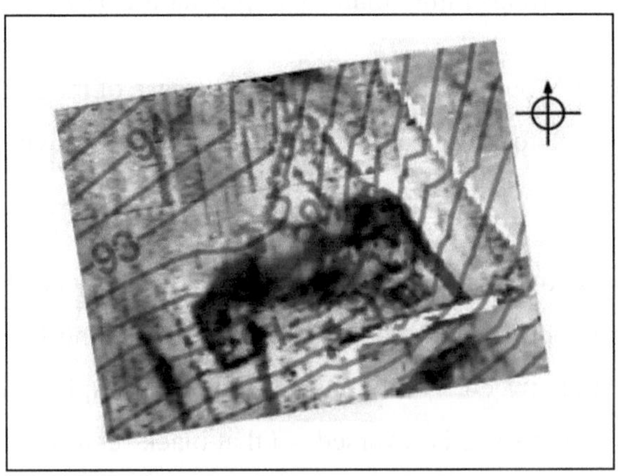

Figure 4.7 An overlay of a res plot of a Roman villa (red) on a mag plot of a ditched enclosure (green). The building lies right over some ditches. Contours are also overlaid. See also Plate 9.

colours. The problem then is whether adding colour enhances the images or just obscures them. Figure 4.6 shows the same geophysics as Figure 4.5 displayed in greyscale and reversed.

Generally, simplest is best, but some colour can help, especially if you have used more than one type of instrument. If you are comparing plots side-by-side or overlaid, it can help if they are different colours, perhaps red for res, green for 'gradi' (mag). An overlay is shown in Figure 4.7 (reproduced here in greyscale, but see also Plate 9 the colour section).

Interpreting the picture is very much one of pattern recognition. However objective the measurements have been, this is a point where you have to be a bit subjective. We are generally good at recognising patterns we have learnt, such as picking out a tune musically, recognising a well-known figure in a cartoon drawing, identifying Orion in the night sky, or finding where we are with a map. Now we have to guess what is under the ground from the pattern drawn on the computer screen. This may sound crude, but understanding this from the start may stop us from making too fanciful claims.

Some patterns may be obvious, such as a ditch found by a magnetometer, or the outline of a building found by res (Figures 4.8 and 4.9) but features can be faint, or natural outcrops may be stronger than the interesting data. There may be features which are obvious but make a pattern you do not recognise. You may need to do wider research to find what structures could make such patterns. Magnetometers may indicate a hearth, but no other feature of a structure. You may need to study the picture closely to observe what is there, and you may need to think how you can strengthen a faint image. Is there another technique which would give a better picture? How much you see

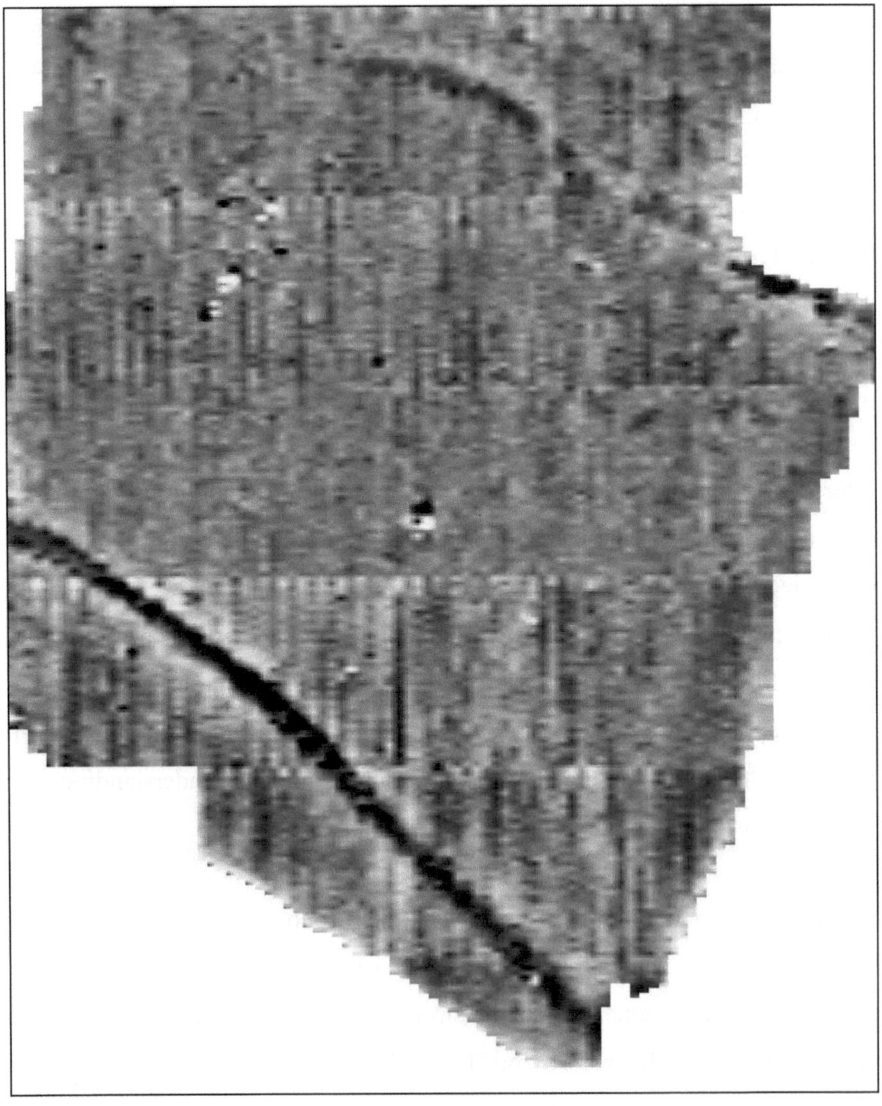

Figure 4.8 Ditch detected by mag. Note also a fainter ditch towards the top of the plot.

Figure 4.9 Part of a building detected by res. Note also high res readings along hedge lines, where stones have been dumped.

can also depend on the amount you have surveyed. It is often easier to pick out features in a large area than it is to try to understand part features which disappear off the edge of the picture, or which cannot be related to their surroundings.

However exciting the display may seem when you first see it, it will not mean anything to anyone else (or even to you in a few months' time), so you need to add some information to help explain it. At the very least, you need to explain where the work was done, what area was covered and what type of instrument was used. Most software has functions to help provide the basic information, but you may want to go beyond that when publishing your work.

First, you need to add a scale and an arrow pointing north, so that you can work out how your survey fits on the map. You also need to add a note saying what instrument has been used and where the site is. These can be added within the software, or should be added in any document that the picture is put into. You may also want to add a scale for the readings. For instance, white to black may represent a scale of 40–200 ohms in res, or –2 to +3 nT in mag. Even if you do not add a scale, it is a good idea to have an idea of the range of the values which contribute. In some cases, there may be much higher values due to interference – for instance, a water pipe may have caused the mag to swing between –100 and +100 nT even if all the archaeology is covered between –3 and +3 nT.

You may want to overlay the geophysics picture on top of a map (providing you do not infringe the mapmaker's copyright). This certainly helps to locate archaeological features within a local area. An example is shown in Figure

4.10. You still need to ensure that there is a north arrow and a scale, even if they are not shown in the example given. You can even go further and overlay results from different instruments on to the same map. This means that you could easily see the relationship of walls and ditches on the same plan. You will need to use some colour in these plots, however, to distinguish between the instruments (an example was shown in Figure 4.7) .

You may even want to add contours to such a map, especially if it covers a large area, say 5 ha, so that you can see the height relationship of the various features. It can also help to know which way water would flow down the ditches you have found. With modern geographical information systems (GIS) software, you might even be able to add the geophysical data to a three-

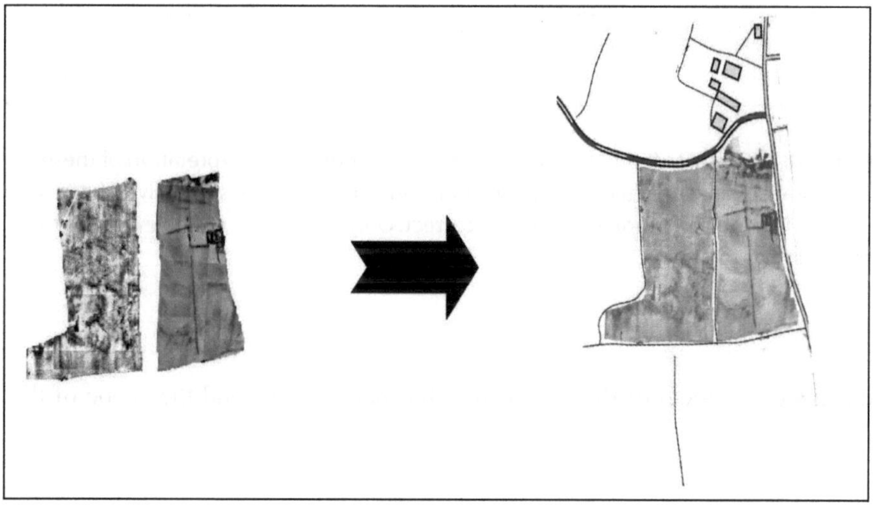

Figure 4.10 Two neighbouring fields, surveyed by res. It is easier to locate the archaeology when they are mounted together on a map.

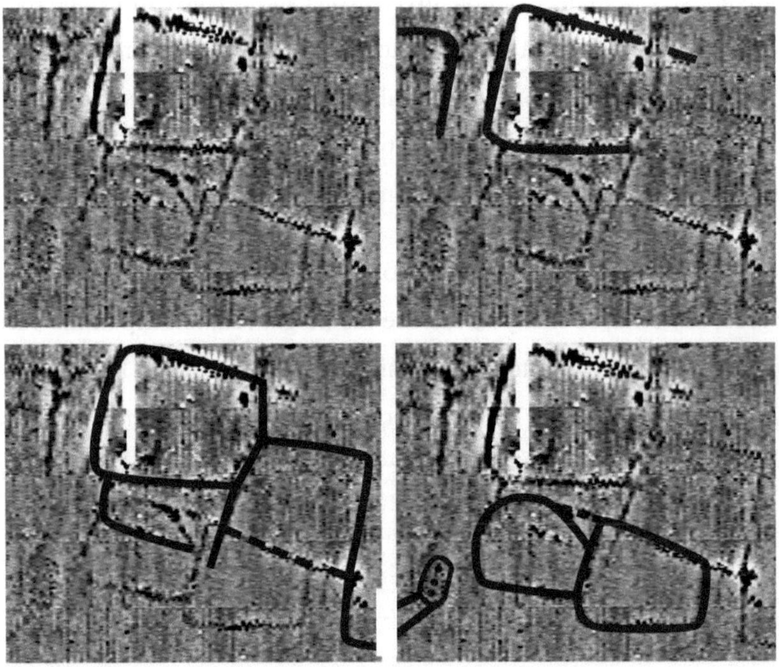

Figure 4.11 Features found by mag in a field, and a possible interpretation of these as separate phases of activity: original plot is top left, then three successive phases of activity. This interpretation need not be correct. Only excavation could reveal the true chronology.

dimensional model of the landscape. But that goes beyond the scope of this book.

You can go further, and try to identify the periods of the various features you pick out, but at this point your display becomes very much your own interpretation and you have to realise that this is subjective. You may wish to trace over the items you think are prehistoric in green and those of the mediaeval period in red, but you have to realise that this is your own best guess, and it is a good idea to show the raw data as well so that others can make their own interpretation. Figure 4.11 shows geophysics output and an interpretation. Do you agree with the interpretation? The one given need not be right.

In the case of other instruments, such as ground radar, the displays you can present may become much more complex, as you now have three-dimensional data. You have to ensure that your presentations are concise and well labelled so other people are not confused by the sheer amount of data. Again, you have to recognise where the display goes from objective data to subjective interpretation. For instance, one form of display is to give a series of 'time slices'. These assume that depth below the surface is just an indicator of age. They may be good for relative age, but if the modern ground surface slopes over the archaeology, this could give a very misleading picture, as the time slice effectively links all points on the survey which are the same depth below the surface. An example of a composite time slice was shown in Figure 3.16 (see the colour section, Plate 6). Pictures like this can be very informative with the addition of depth information. It is important, however, to know what you are displaying and the limitations of your display technique.

The more common processing software packages are discussed in more detail in Appendix B.

This concludes our look at the technical and instrumentation side of conducting a geophysical survey. In the next chapter, we look in detail at the practicalities of how to conduct your own survey.

[illegible]

[illegible]

[illegible] Appendix B.

[illegible] practicalities of how to conduct your own survey.

5

Geophysics Survey Campaign

5.1 Preliminaries

The first thing you need to know is whether such a survey is legal, and the rules will vary from country to country. The law of the land may permit survey, but only with special permission from the authorities. Even where access laws are liberal, you still may need special permission to work on some sites which may be protected by law. Where sites are scheduled as ancient monuments, not only is excavation prohibited without special licence, but so is survey. Licences for survey may be quite easy to obtain, but you must check first before starting any campaign. In the first instance, you will need to find the local or national authority's official archaeologist to check the rules. This is a time when internet searching, for instance county heritage environment records (HER), will do much of the initial work.

There may be extra legal requirements, such as the need to have insurance to cover other parties. You will certainly want to know that the equipment you use is insured as its value could possibly be fairly high .

The major task then is to find a suitable site to investigate. This will depend on your personal preferences, whether you are interested in prehistoric or First World War periods or anything in between, and will have some bearing on the locality that you choose. Alternatively, if you want to stay in one area, that may restrict the types of site which can be found.

You then need to decide whether you are intending a large area survey to understand the features in the landscape or whether you are interested in concentrating the survey in a small area, perhaps indicated by visible earthworks.

Earthworks are one way of showing that man has been active in past times. Another is use of place names. A settlement name in the Old World can often be an indicator of earlier settlement, perhaps suggesting a major Roman site, or the presence of monastic works. However, the settlement probably covers an area larger than you can contemplate surveying, so you need to localise to a smaller area, perhaps about two hectares. If a field is small, just a few hectares, and has a name, this may give a clue. For instance, Blacklands, Three House Field, Stanchester or Old Way Paddock are names which suggest past use of that ground. The field's location also has to be known, so an old map is more use than an old document which just lists names. There may even be antiquarian writings describing finds at a particular place. The names are not in themselves sufficient proof that activity will be discovered.

You might surmise that a field is worth investigating from a modern map or aerial photograph, for instance, if the road makes a detour around it, many lanes converge on it, or field boundaries make a sudden sharp zig-zag. Knowing the lie of the land may be a help. A sheltered point close to a spring but above the flood plain, and commanding extensive lands may be the most likely point for earlier generations to have had a settlement.

Is anything visible on the surface of the field to indicate previous activity? Earthworks may not be obvious, but there may be a scatter of artefacts, a patch of very dark soil, or areas where the crop does not grow well, and all these suggest past activities. These signs are more likely to show in arable fields, whereas those that have been under grass for a long time may give little indication of the remains that they are concealing.

Once you have located a site which you want to investigate, it is most important to obtain the permission of the owner and of the tenant of the field to enter it and investigate it further. Their attitudes may vary from delight at your interest, through indifference but willing to let you in, through concern to outright hostility. If you meet hostility, it is best to look for another site. In all cases, you need to think about what the land is being used for, and whether there are times of year when your presence would be very inconvenient. For instance, demanding to survey while crops are being harvested is likely to generate hostility, while surveying a grass field during the winter while the cattle are inside may be much more acceptable to the farmer.

I have concentrated here on rural sites, and when working those it is a good idea to gain at least a rudimentary understanding of farming techniques. This helps you to avoid irritating the farmer, and may also give you clues about how the field has been used in the past. Surveying in urban sites is possible, but there is a much greater likelihood that the ground has been disturbed and this will obscure the archaeology. These would not be good sites to begin on, although they may yield good data to experienced geophysics practicioners.

5.2 Geography, geology and weather

Choosing the right time of year for a survey can affect the survey results as well as meeting farming requirements. Extremes of heat, cold and wet are best avoided, both for the sake of the equipment and the operators.

Winter may be a better time for survey than summer, even allowing for a shorter working day. Obviously, deep snow or extreme frost preclude winter geophysics campaigns, as do the very short days of high latitudes. However, crop or grass growth is very slow, so less harm can be done by walking on it. The vegetation of any wasteland or uncultivated ground will also be at its lowest. Many animals will be kept indoors at this time of year, making the fields much more accessible. If it is not so wet that the ground is saturated (this means that in resistance measurement, stones are 'short circuited' by the ground water, so are less distinctive), it is often easier to work in muddy conditions than on hard baked summer soils, where getting good electrical contacts can be problematic.

If benign conditions extend into spring, then a little more warmth and longer days can offer more comfort than winter, but once the grass starts growing, you may find farmers want you off the land.

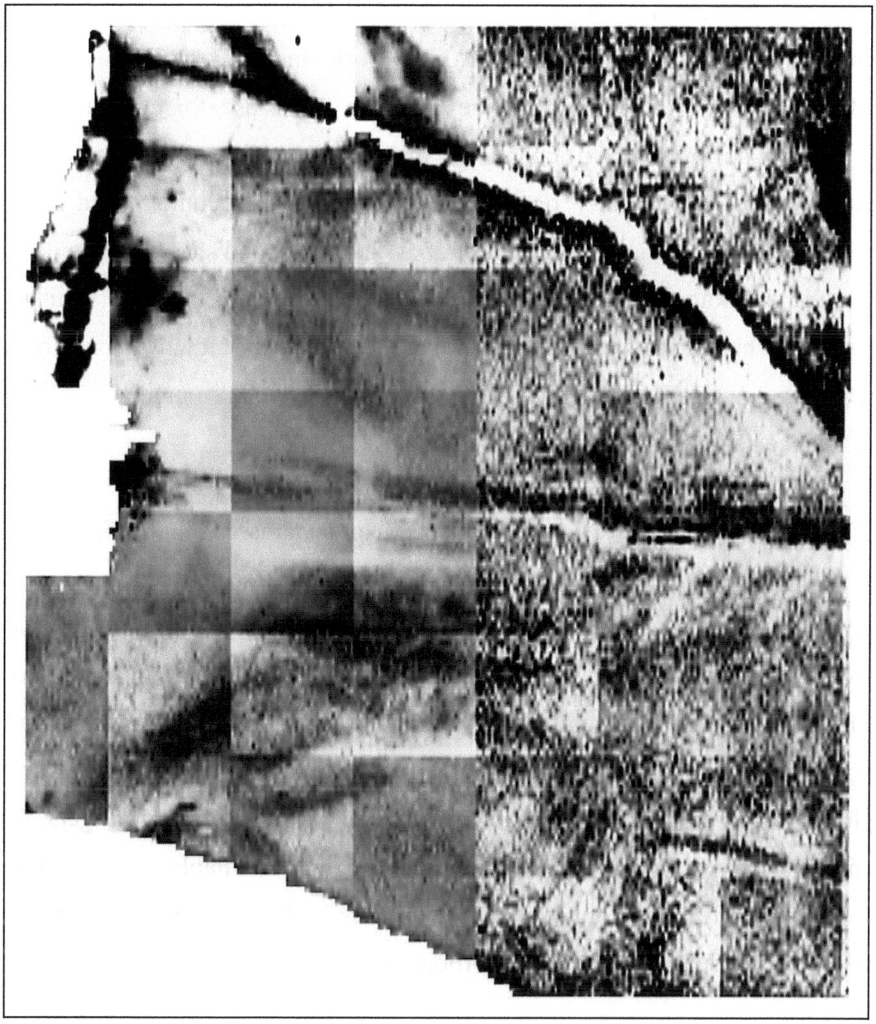

Figure 5.1 The difference between summer and winter in resistance. The left side was surveyed in winter, the saturated ground masking the response to stone. The portion on the right was surveyed in a hot summer, with the clay baked hard and cracked.

Summer and autumn offer hotter weather and longer days, but they can be uncomfortably hot for sustaining a day's work. Insects and flies may also add to discomfort. Fields are more likely to be in use for cultivation or grazing, making access difficult. There may also be problems with resistance measurement, particularly on clay soils. These can become baked hard, making it very difficult to insert probes and get good electrical contact. Ground shrinkage and concentration of the remaining moisture may result in patches of high and low resistance, giving the plot a very blotchy appearance, which can mask any archaeology. The difference between summer and winter resistance surveys can be seen in Figure 5.1.

Whatever season you work in, you need to be dressed appropriately. In summer, this could mean being covered up sufficiently so you can spend long periods in bright sun without risk of sunburn. You may want to have sun block creams to hand. In winter, you need really warm clothing. It is not sufficient to dress for a short walk, you will be out in the weather for several hours, so you will need a heavy coat, hat to retain head warmth, boots, and gloves which are warm but sufficiently thin to allow use of your fingers. At any time of year, it is wise to have plenty of fluid available. You can feel dehydrated after a day out in winter as well as in the summer sun. Clothing will be discussed in more detail in Appendix C.

Some basic geography must be applied to any campaign. This includes the historic political geography of your chosen region, so that you know what sort of habitation and human activity to look for. You also need to look at the physical landscape to see where settlements were most likely. There may have been changes in climate and land use since the period you are interested in, but the structure of rivers, flood plains, terraces, steep hillsides and springs will not have changed greatly and good conditions for living were even more important in past times than they are now.

There have been some changes since ancient times, though, particularly migration of soil down slopes, and this can have an effect on your survey even if you have located a settlement site correctly. The extra depth of soil at the base of hill slopes may be sufficient to bury the archaeology deeper than your instruments can detect. Settlement activity can extend up steep slopes, so there may be archaeology to find. There may come a point, however, where it becomes too steep for safe surveying.

Geology can have a massive effect on the success of your survey, and this applies to both mag and res. There has to be sufficient material of magnetic influence for the mag to detect anomalies in it. This means that there needs to be plenty of iron present in the soil and underlying rocks. Fortunately, such rocks advertise themselves by strong red or yellow colouring. Surveys are less likely to be successful on grey limestone or white chalk, although a capping of clay over the latter provides good magnetic conditions. Magnetic survey may be good even on unpromising geology if lots of domestic and farm waste has been spread over the ground, but only that small area may give good results.

Res works well in stony areas whether or not the stone is yellow. However, the stone must be a useful building material, so that it has been historically easier to build with stone than with wood or clay. In areas of unsuitable geology, building remains may only be collapsed mud or rotted timber, and there may be nothing for res to detect. Indeed, such structures may be very difficult to detect at all, unless mag can recognise a hearth.

These are guidelines to help you select a suitable site. In the end, you have to do the survey, and be prepared to be equally delighted or disappointed with the results. Providing the survey has been well done, even negative results make a statement about the archaeology of a site.

From now on, I assume you have identified a suitable site and obtained the necessary permissions from owner and tenant and, if needed, from local and national authorities. Then you are ready to start on the survey.

5.3 Setting up for surveying

It is possible to do a survey solo, but it is much easier having a team to help with the work. That may sound a bit like cheating if the survey is part of a degree dissertation, but it is not. Learning to run a team is important in geophysics. You have to collect together a team which may include fellow students ('If you help me with my project, I will help you with yours.'), or friends with no previous archaeological knowledge. The landowner or tenant may even be interested enough to want a try. It is possible to be a solo user of either instrument, but that is certainly not recommended for beginners. You need at least two for res, although three makes for light work. You need three for comfort for mag. If you have enough people, you can run two teams at once, on different instruments.

Once you have collected a team, they need to know where to assemble, where they can park cars safely, what to wear and what to bring for food and drink. You need to get one car close enough to the site that you can carry the equipment comfortably, but park far enough away (at least 30 metres) that it does not interfere with a mag survey.

If you have access to geophysical survey instruments, you should also have access to the ancillary kit. Here are the ancillaries you will need to make up grids: base lines (minimum, 4), strings to walk along (minimum, 2), grid marker poles (minimum, 20), plastic tent pegs (minimum, 10), measuring tapes (minimum, 2). The tapes need to be at least 30m long, preferably 100m.

You will need to decide on details of your grid, what area to cover and how to divide it into grid squares. In the example here, we will assume 20m squares. Some professional groups may walk very long lines and make large grids, but that is definitely not for beginners. The base lines are usually made up from washing lines. They need to be just over 20m long, with 21 marks (coloured tape) each one metre apart. The ends are looped to hold a plastic tent peg. The grid marker poles can be bamboo sticks or dowel rod, typically 400mm long. If you can paint or tape them with gaudy colours, that helps to make them visible. You will need strings to walk along to help you to keep an accurate path. These also need to be just over 20m long with ends looped to hold pegs. They should be marked every metre at least, and you may want to mark them at smaller intervals if you aim to do finer measurements. They can be made of washing line or from pre-shrunk sash cord with the marks sown through. Marks need to be bright to be visible. These are discussed in more detail in Appendix C.

It helps if you can align your grid on magnetic north, but it is not essential if your chosen field has a boundary which provides an obvious alignment.

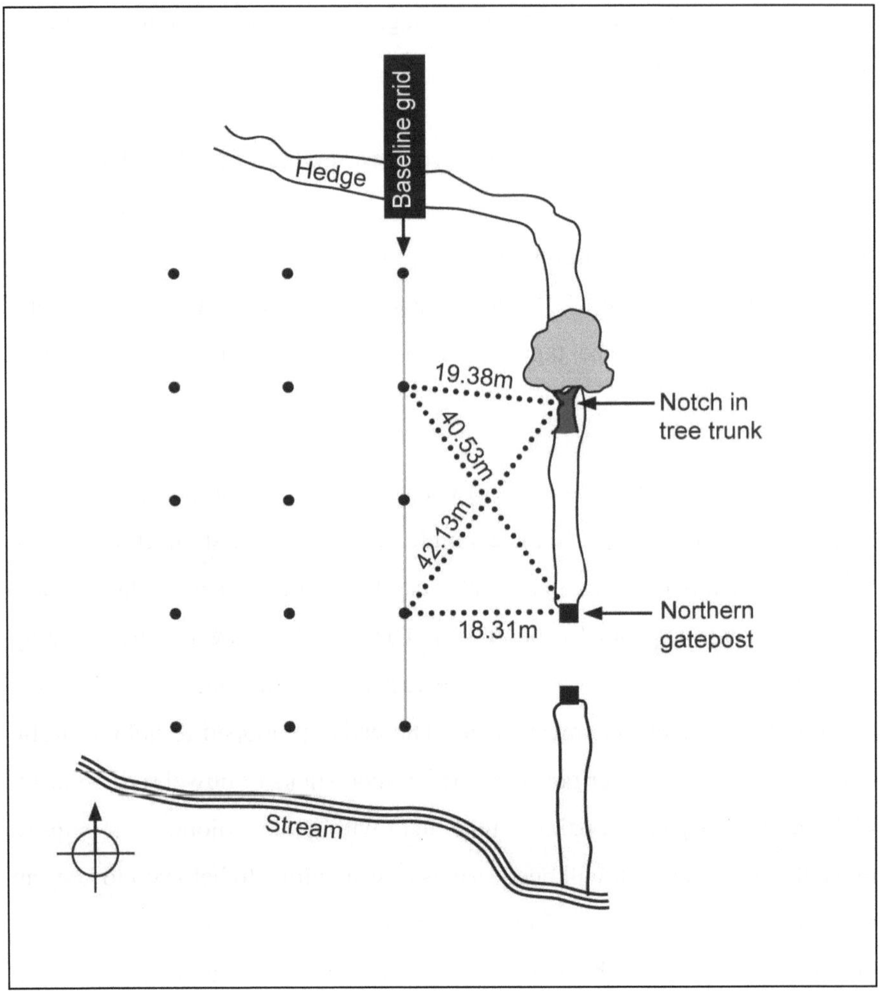

Figure 5.2 Make a sketch of the field and of your grid. Measure the position of two grid posts each to two fixed positions which will be identifiable in the future. You can then rebuild the grid later if you need to.

The first thing you need to do is to establish a straight line of marker poles as the base of your grid. The marker poles need to be at 20m intervals. It is a good idea to make some measurements from two points on this base line each to two points in the field which can be identified later. These might be a tree, a fence post or a gate post. They will not be there forever, but they are likely to remain some time, and this means that you can measure these two points back into position and then re-form the base line. You can then rebuild the grid if you ever need to go back to repeat the survey or re-survey with different instruments.

You need to keep the measurements, best in a rough sketch (Figure 5.2). You could also photograph your fixed points to help you relocate them. It is not sufficient to measure the position of the two points with a hand-held GPS as these are not precise enough to establish exact positions. You also need to keep a sketch of the grids you lay out. You will need to add to this sketch as you add more grids. Keep some copies of this sketch so that you can number the grids as you survey each in turn. You will need a separate sheet with grid numbers for each instrument, as you may not measure grids in the same sequence. An example sheet is shown in Figure 5.3.

Next, you have to turn the base line into a series of squares by constructing right angles. One way is to use a right-angle prism sometimes known as an 'optical square'. Two prisms or mirrors give a view at right angles to the direction you point it, and there is a space between these to see where you are pointing directly. If you stand over one of your grid poles and look across the line, the lines of poles either side will be visible as images in the prisms, and when the images are aligned, you are pointing exactly at right angles to the line. You then need someone to hold a pole directly in front of you, say 40m

SITE: Normeads METHOD: Resistance

		1	2							
		3	4	5	6	7	8	9		
		10	11	12	13	14	15	16	17	
	21	22	23	24	25	26	18	19	20	

Figure 5.3 Use a squared sheet to number the grids as you complete them. This will help you remember the sequence for data processing.

away, and you need to direct them right or left until their pole aligns exactly with the images of poles to either side. The line to their pole is then exactly at right angles to your base line. This is shown in Figure 5.4.

If you do not have an optical square, you can use tapes to construct a right angle. This is where you have to use Pythagoras' theorem. It is not the mathematics but the results that count. Once you have two sides of a right-angle triangle, you know what the third will be. If you have two sides of 20m with a right angle between them, then the diagonal will be 28.28m. For the mathematical that is $20^2 * 20^2 = 28.28^2$, or $28.28 = 20 * \sqrt{2}$. Fortunately,

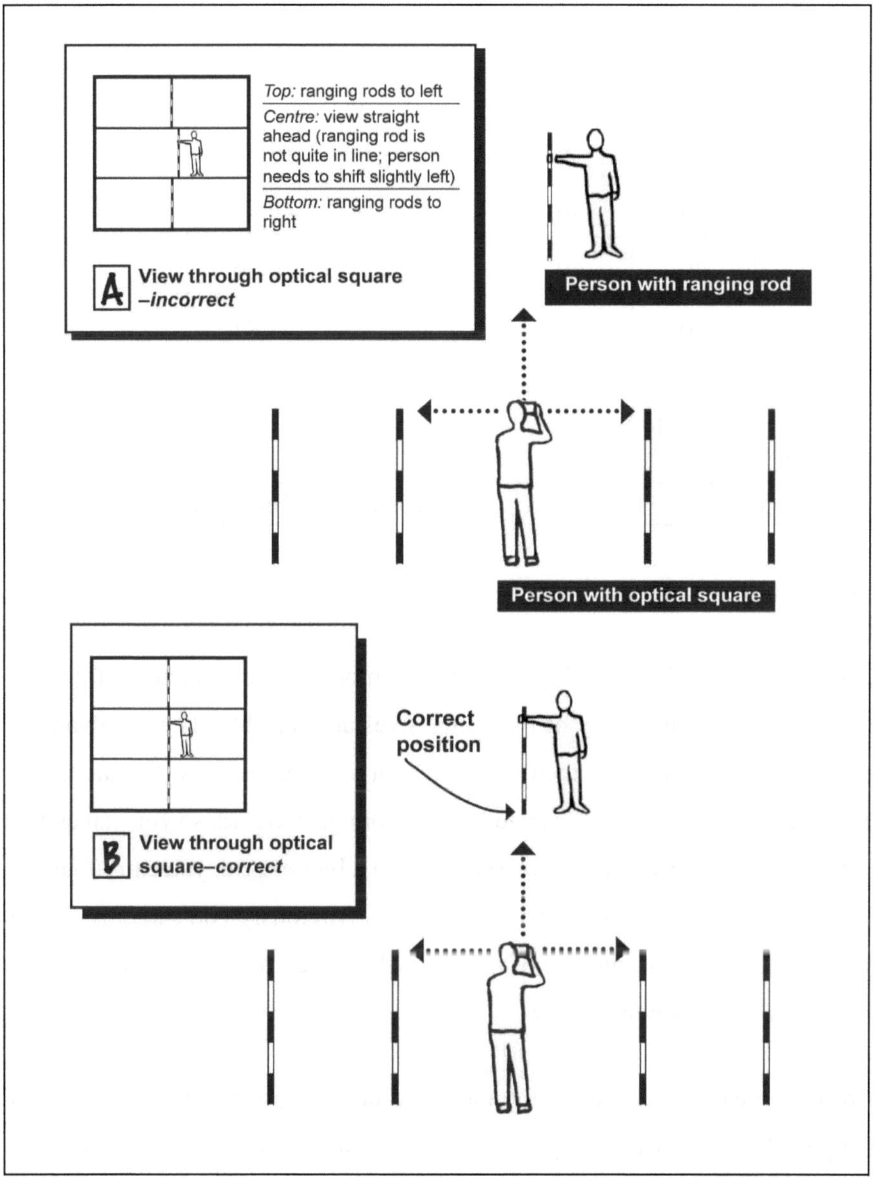

Figure 5.4 Using an optical square to set up a right angle.

the numbers are quite easy to memorise, so you can build up squares with differing side lengths:

Sides	10	10	diagonal	14.14
	20	20		28.28
	40	40		56.56
	60	60		84.84

A useful rectangle to know is

Sides	60	80	diagonal	100

These are shown in Figure 5.5.

If you only have 30m or 50m tapes, you can only lay out squares of 20m. If you have 100m tapes, any of the above sizes are possible. The disadvantage of laying out a 20m grid, then one next to it and another next to that, is that any errors in your squares keep adding up from grid to grid, so you can end up with a badly distorted grid. If you can lay out a bigger grid, you can subdivide it into 20m squares without increasing errors. Obviously you can only do this well if you can see to the end of the line, which may not be possible over the brow of a hill, or on misty days.

Errors can easily build up if the tapes are not straight and taut, which may happen if the wind is strong or if they get caught round weeds or stubble. Pulling a tape too hard can also pull the pegged end out of the ground so that it is leaning 5cm towards you. Then a measured length of 20m is actually 20.05m. This can result in right angles being out of true by up to half a

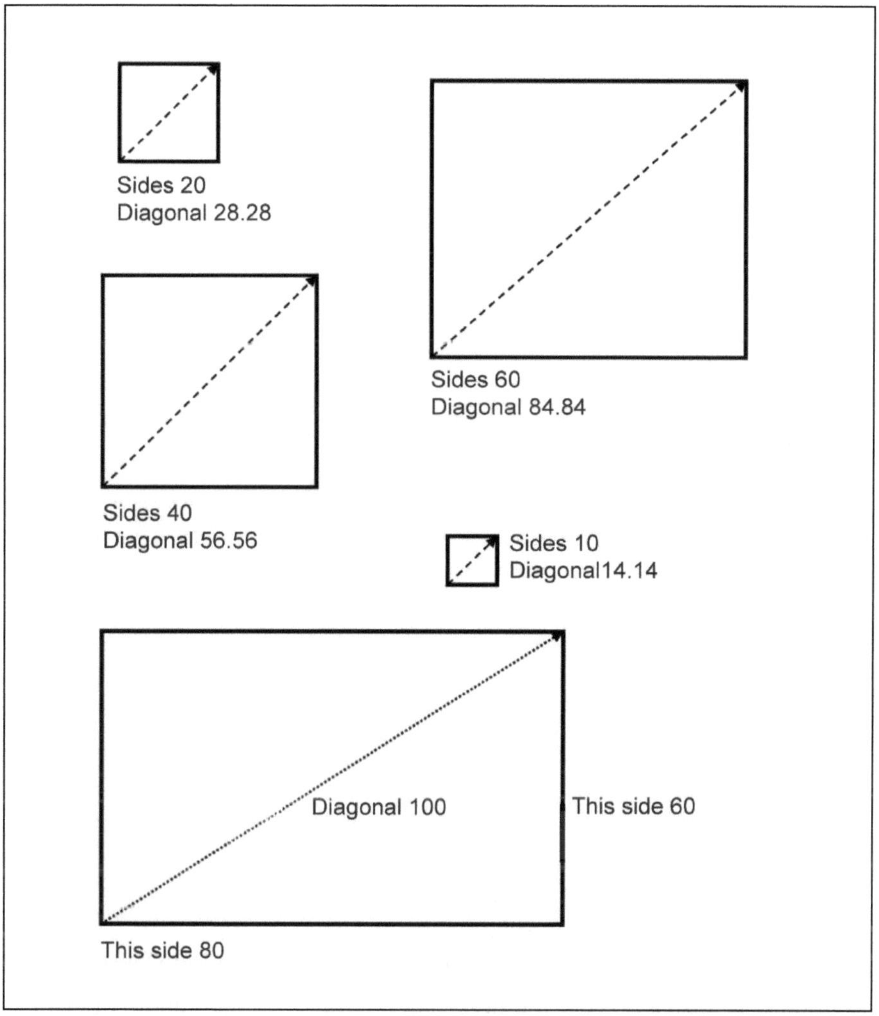

Figure 5.5 Constructing right angles using tapes.

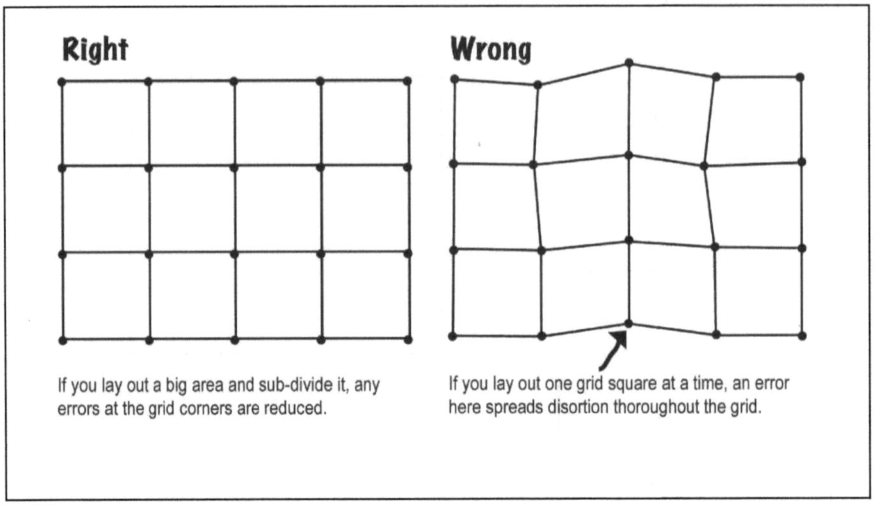

Figure 5.6 The grid gets distorted by one wrong measurement, and will get more distorted as it is expanded. The larger the area you can set up in one block, the less likelihood of an error growing.

degree. This is not large, but it mounts up as you add grid after grid. If you have 60.05m instead of 60, the percentage error is smaller, and that error is spread over more grids (Figure 5.6). Figure 5.7 illustrates some of the ways that errors occur.

After you have created some grids, you can lay out one base line east–west between two corners, a second base line east–west between the two corners to the north, and walking lines north–south between the base lines. However,

a

b

c

d

Figure 5.7 Ways that errors occur: (a) tape bowed out by the wind; (b) tape caught round a tussock of grass; (c) peg and tape end pulled several cm from the grid corner; (d) lines not pulled straight.

there is another matter to decide before you do that. Remember that your 20m base lines have 21 markers, so which 20 of those will you use? You need to arrange it so that grids just touch each other but without overlap. If you start on the first line of the grid (between the corner markers), you will finish just short of the end of the grid, so that the next line, the one between the corner markers, becomes the first line of the next grid. Alternatively, you could start 1m in from the line between the corner markers, and finish going down the line between the next pair of corner markers. These options are shown in Figure 5.8. Once you have laid out base lines and walking lines, remember that these are trip hazards. You need to ensure that everybody knows where they are to avoid tripping on them. This includes your team members and any casual passers by.

The same arguments also apply to walking up a string. You can either start on the base line and finish one mark short at the top, or you can start one mark up and finish on the top line, as shown in Figure 5.9. If you really wanted to, you could start and finish half a mark in from base lines and corner posts, but it will be more difficult to see where to take measurements.

What matters most is that you have a consistent operation across whatever instruments you use. You will also need to decide your starting point, the direction you walk and the direction of the base lines. It is good policy to keep these standard for all surveys and for all instruments. Indeed, most processing software packages demand that you keep starting point and direction consistent throughout a survey.

You may have already been taught a grid layout and want to continue with that. In this book, we will assume that base lines are laid east–west, with

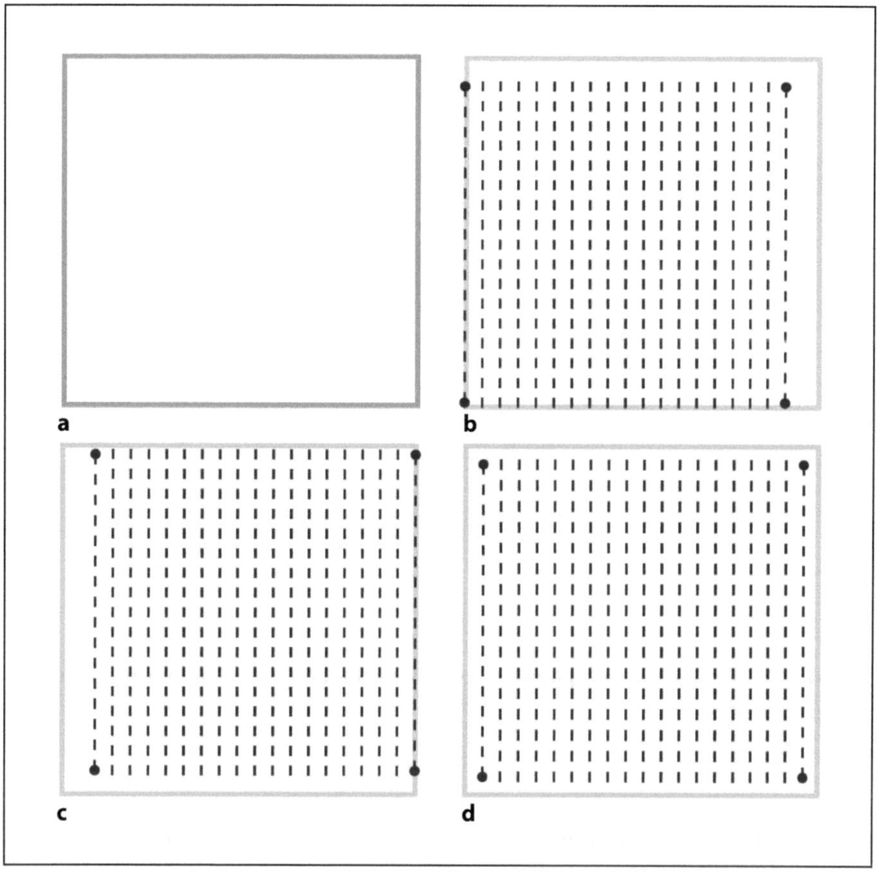

Figure 5.8 Possible ways of laying lines in a 20m grid: (a) the blank grid; (b) start at SW corner, finish before N line, finish before E line; (c) start N and E of SW corner. Go to N line. Finish on E line; (d) Start inside square and leave an equal gap from all edges.

strings for walking laid north–south. Start in the south-west corner, with the first line 1m east of the western corners (so that the final line is between eastern corners), starting one mark up from the southern base line, so as to finish on the base line. This was shown in Figure 5.8(c). The first walking

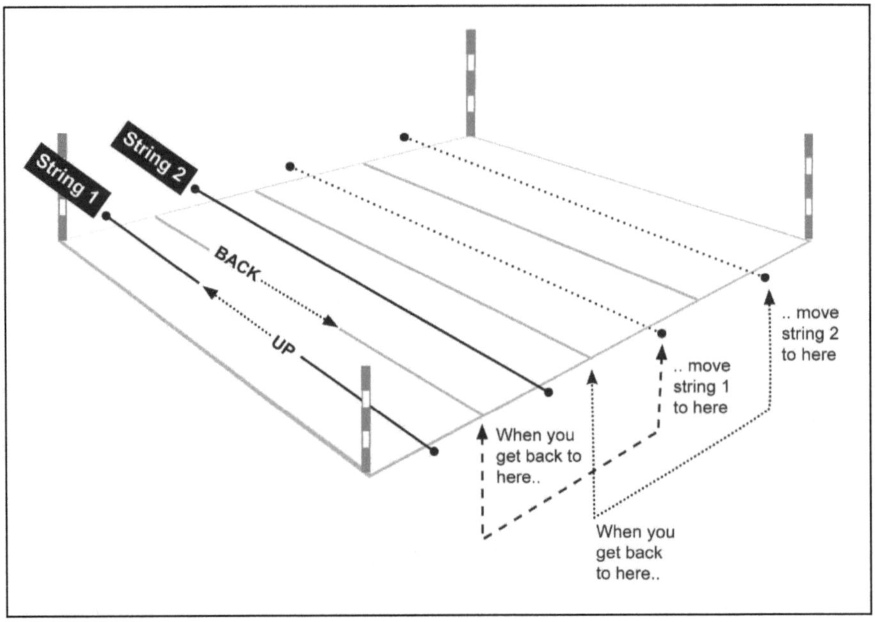

Figure 5.9 Laying out walking ropes, and walking the grid, based on the arrangement in Figure 5.8 (c).

line will then be set running from 1m east of the southwest corner to 1m east of the northwest corner. The normal procedure is to walk up the line, and walk down the gap between lines, so the second line is placed parallel to the first, but 2m east. Any more lines are placed a further 2m east again. This is shown in Figure 5.9.

We have assumed grid squares of 20m, which is a convenient size when learning. You could lay out 30m grids, but they have over twice the area, 225% the number of readings and take about twice as long to do. Professionals may use even larger grids, but that is not recommended while learning. Twenty metre grids have another advantage in that they can cover a field quite close to the edges. No grid layout will ever match a field exactly, so there will always be some partial grids around the perimeter. You can decide whether you want to measure these partial grids so that you have coverage of the field right up to the edge, or whether you will just survey the main grids. Most equipment has facilities for measuring only part lines, or for putting in blank lines where the grid runs off the edge of the field.

Once the instrument operator has gone up a walking line and down the gap between the two lines, the first, westernmost, line is then 'leapfrogged' over the others until it is 2m beyond the easternmost line. This way, the north–south path to walk keeps progressing in an easterly direction. It is then best to have another pair of base lines set up in the next east grid square, so that you can continue smoothly with the survey after you have finished the first grid.

If you also decide to carry out a metal detector survey, you do not need the walking lines, as the detector is swept in a wide arc, covering about 2m. However, it is good policy to sweep within each 20m grid square at a time and record which square any finds come from. Then you may be able to correlate any detector finds with features such as buildings. Greater precision than the 20m square is not usually relevant as detector finds will most likely have been moved around by the plough. You will need to check each country's law on ownership of any finds.

If you are working on plough soil, you may want to collect any visible surface finds. Again, it is a good idea to keep finds from each 20m grid separately identified, in case you can relate find type or find numbers to particular features which the geophysics reveals.

5.4 Resistance measurement

We consider res first because it is less demanding, albeit it is harder work physically than mag. We will also assume use of the twin probe configuration, which uses two probes on the frame and two remote probes connected to the frame by a long cable. The frame also holds the measurement box and data logger. The science of the measurement was considered in Chapter 2, and practical instruments in Chapter 3. In this chapter we discuss how to use the instrument. Exact details of which button to press cannot be stated as the instruments differ from one maker to another. The common instruments will be described in detail in Appendix A.

The twin probe arrangement may look awkward with a long cable getting in the way, but the results are easier to interpret, and it keeps the frame simple and lightweight. It is much easier to have to try to get only two probes in the ground together, not four. We will also assume here that the frame has 0.5m spacing between probes. Frames with 1m spacing can be obtained, and may provide greater ground penetration, but they are heavier and more cumbersome.

The control box and data logger need first to be fixed to the frame, and the frame probes connected (see Figure 5.10 and Appendices A1 and A2). The cable to the remote probes must also be connected to the box. The two connectors should be labelled, and may be different, so that each will only connect to the right place. It is a good idea to tie the remote probe cable loosely round the frame first before making the connection. This acts as a strain relief, so if there is a sudden tug from the cable, it does not damage the connector.

The remote probes do have to be remote. The physics assumes that the spacing between the frame probes is negligible compared to the distance between the frame and the remote probe pairs. As a rule of thumb, the minimum distance should be 30 times the distance between the probes on the frame. This means that the remote probes must be at least 15m from the nearest point on the grid. If the cable to the remote probes is 50m, you can place the remote probes so that you can complete two grid squares without moving the remote probes (Figure 5.11). You must pay out all the cable, and you must not let it form a tight coil at any time, as this can set up inductive effects (see Section 2.1).

Figure 5.10 Connect the frame probe and remote probe cables to the meter. Make sure there is a loop in the remote probe cable, so that takes the strain, not the connector.

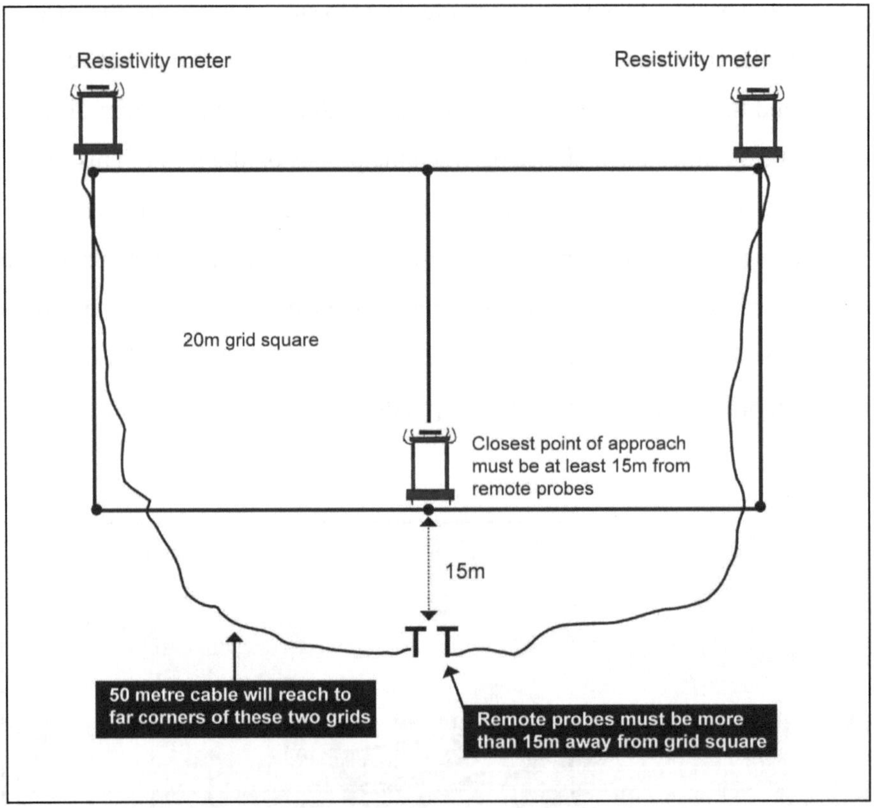

Figure 5.11 Setting the remote probes position to do two squares without needing to move them.

Set the frame up close to the point where you want to start. The reading that appears will depend very much on the soil. Wet clay soils can produce readings as low as 10 ohms, while sandy soils may give readings of hundreds of ohms. In the latter case, you find you will get no reading until you change range on the meter. In general, use the lowest range you can to get your readings as precise as possible, but note that a 'signal' may produce a reading up to 10 times the normal reading. For instance, in a clay field, a wall may give readings of 80 ohms where the soil gives just 15 ohms. On drier soil, a wall may be 400 ohms where the soil is 200 ohms.

Once you have a reading, move the frame about a bit over a few metres to get an idea of the range of readings you expect. If the reading changes greatly, you may be just on the edge of a feature, and you should find an area of lower readings to set up. You might place the remote probes about half a metre apart and work with whatever reading that gives, or you may want to start at a precise value, for instance, 25.0 ohms. Moving the remote probes further apart lowers the apparent resistance value. Moving the remote probes closer together increases the apparent resistance, but the probes should be kept at least 0.3m apart.

One advantage of res is that it is easy to make it trigger automatically. Putting the probes into the ground takes electrical current from zero to a measurable value. The machine can be made to trigger and take a reading once the current reaches a particular value. If that value is too low, there is a risk of false triggering producing poor results. Normally instruments work once the current is high enough to measure properly, but they may have an extra setting which takes a measurement as soon as current flows. You still have to get the probes into the ground, and if the current probe on the frame hits a stone, it may not be possible to get a reading until you have found a patch of soil to get the probes into. The stone will cause a higher reading, but will not then stop the reading. If the voltage probe hits a stone, you may get a reading, but it can be very high.

The advantage of triggering on a definite signal also provides another bonus. The instrument will only take another reading after it has been pulled out the ground and then reinserted. If you need to stop the survey for any reason, either to move cables or lines or even just to take a rest, you can leave the instrument planted in the ground after its last measurement and it will sit there while you do what you need to, and just pick up again after you lift it out of the ground and put it in at the next point.

Check that the logging system is correctly set so that it takes the right number of readings per line and the right number of lines. If it has an averaging system for working over different ground conditions, choose the shortest time setting you can. You normally only need longer averaging times over very disturbed ground, such as old building sites. Using a longer averaging time will slow you down considerably.

Put the frame in the ground at the first point to measure. Start the logging system. You will then need to log the first reading manually. After that, lift the frame out of the ground and put it back in at the next mark on the walking line. It will automatically log the reading when it has made good enough electrical contact. Do not push it further into the ground than you need; that saves effort in pulling it out again. The instrument will take a while to take a reading and will indicate this by bleeping for the end of the averaging time. Keep the frame in the ground until the reading has been taken.

The instrument will also indicate when it has reached the end of the line. You must make sure you turn round and start the line back at that point. If you have accidentally taken a reading twice or missed a point, you must still get the end of line signal and turn round when the logging system expects you

Figure 5.12 At the end of a line, reverse in an arc, keeping the res in front of you. This avoids tangling with the cable.

to. If lines walked and lines logged get out of sync, that is the worst possible condition for producing a coherent picture. When you turn round, it is best to reverse round in an arc, as in Figure 5.12. This leaves the cable in front of you, so you can adjust it without tangling it.

Walk back between the two walking lines, taking a measurement as you pass each mark on one line, until you reach the end of that line, when again the instrument will tell you it is the end of the line and you must turn round and head up the next line. Continue in this fashion until you reach the end of the grid, and the instrument also tells you that the grid is finished.

If you are surveying a partial grid, the res has controls to help you limit the logging to where you can take real readings. If you are going up a line, but it is, for example, only 12m before it hits the hedge, you will need to take readings up to the hedge and then put in dummy readings to the end of the line, and then dummy readings back down to where the next line would emerge from the hedge. There will be a button for inserting dummy readings. It would be tedious to put in two lines of 8m of dummy readings, so instead of repeatedly pressing the dummy button, there are other buttons. One will put in dummies up to the end of the line. One will also put in dummies for the following line, down to the point opposite where you have stopped and added dummies. The exact button sequence cannot be given here as it differs for different instruments. Using these facilities makes it easy to survey partial grids.

You can also have a partial grid which has full length lines but not the full number of lines. If the last three lines of the grid are beyond a hedge, you need to survey as much of the grid as you can, then press the button which completes lines with dummies until you reach the end of the grid. However, if it is the first four lines of the grid which are missing, you need to put in the dummies before you start surveying, again using the facility which fills lines to the end with dummies. You must put the dummies in at the correct missing line numbers, or you will end up with discontinuities when you download your survey.

It is best if res can be done by a team of three. As well as the res operator, it helps if someone walks beside them carrying the cable. This is not essential if you are surveying over short grass, but the cable will snag very easily over rough ground or stubble, bringing your progress to an abrupt and uncomfortable halt while you untangle it. If it has got tangled some distance off, you would have to leave the line to untangle it, and there is a risk that you will come back to the wrong point when restarting. Where a second person holds the cable, it is their job to make sure it runs freely and does not snag.

The second person also moves the walking line for you while you are turning round back at your base line. The third person stays at the top end of the grid and moves the other end of the walking line. The two then leapfrog the line over the one you are about to walk up and peg it in the ground 2m beyond, so that it is ready for you to walk up next time. Make sure the walking line is always moved under the cable. If you get one end over the cable, you will tie a knot which has to be untangled before you can move on.

With a team of three, you can rotate each task at the end of each grid, or even mid grid, so that each person has a turn at the tiring work and then some rest moving the top ends of the walking lines. If there are only two of you, the most important secondary job is to ensure that the cable does not snag, so they should walk with you. However, when it is time to move the cable, you have to plant the res in the ground (where it will take the next reading and then sit quiet) while you move that end of the walking line. The cable carrier has to race up the line and move the top end while you move the near end. If you are working solo, you have to plant the res in the ground, move the near end of the walking line, and then go up to the top of the grid and move the far end. This makes for a lot more walking and also uses up a lot of time. Having at least one companion is faster as well as easier and more enjoyable.

5.5 Magnetometer survey

It is easy to set up an automatic trigger for the res, but that is not the case with mag. The mag is sampling the magnetic field around it the whole time, looking for minute changes, so there is no sudden change to act as a trigger. You either have to trigger it by hand or by using a timer. Most surveyors learn to use the automatic timer, but this means that you must walk at a constant pace, even over rough ground. It is easier to start using a manual trigger. This will be a button on the instrument or a hand-held box plugged into the mag.

Remember that the mag is sampling the field all around it, including you and your team. If they have any magnetic items on them, the mag will detect them, not what is in the ground. Items can include keys, loose change and any clothing with studs, buckles, eyelets and zips. For ladies, underwired bras are the biggest problem. You need to dress non-magnetically and put any magnetic items somewhere safe away from the survey area. Anybody who has not removed all magnetic material should stay at least 2m from the magnetometer. This is considered in more detail in Appendix C. It is usually possible to wear glasses while using the mag, but even they must have non-magnetic frames for calibrating.

Magnetometers take a while to settle as they need to reach a constant temperature. It is best to set the mag up and switch it on before any other operation, so that it has at least 15 minutes to settle. Once it is settled, you need to check that all those who are going to handle it are non-magnetic. First Make sure that it is already well away from any magnetic interference. Then check that as you go to pick it up, the reading does not change by more than 1 nT. If you can hold it without upsetting it, make sure all your team can come up close without affecting it.

Note that each of you will have a slightly different effect on the mag reading. This means that it is essential that only one operator uses it during the survey of a grid. You cannot pass the mag on to another person part way through a grid.

Before you can start a grid, you have to calibrate the instrument. This is an automatic procedure on Bartington mags and a manual procedure on Geoscan mags. The concept is the same in both cases, so the Geoscan procedure is given here to illustrate the method. The two fluxgates need to be electrically balanced but they also need to be aligned correctly. The tube of the Geoscan devices has two caps which need to be removed, each revealing a fine adjustment screw. There is a point on the head where you can insert a small tool, a non-magnetic screwdriver (there is a tool in the case), to adjust the electrical balance.

You will need to know north, south, east and west directions at a point well away from any signals. If you have set your grid up north–south, you can use that alignment. Otherwise, you have to determine north with a compass and set a line on the ground. Make sure you remove the compass!

Hold the mag with the head pointing north and note the reading. Turn the instrument around until it is pointing south. It should give the same reading (to the nearest 1 nT). If not, note the reading and turn it back north. Reach down and adjust the lower fine adjustment screw until the reading has moved half way to the south reading. Then, when you turn the mag south again, the readings should be the same. Keep doing this until you get the same reading both ways round. You have then adjusted the lower fluxgate so that it is pointing exactly downwards.

Next, point the head to the west and note the reading. Turn the head to the east and note the reading. If it is different, point it back to the west and adjust the upper fine adjustment screw until you are midway between the two readings. Turn it east again and check. When it is the same both ways, the upper fluxgate is exactly vertical. Recheck the north–south setting. You should then press the 'zero' button (wait for four beeps) to set zero.

Next you have to check the balance between the two fluxgates. Keep the head pointing north. Swing the mag up until it is upside down. It is easiest to do this using an underarm grip. You must keep the head pointing north. You then need to take a reading while it is upside down. This means crouching down under it on the older FM36, but the newer FM256 has a means to take a reading and hold it. Then you turn it back to normal and note the difference between the two readings. Use the tool to adjust the reading until it is midway between the two readings. When you repeat the upside-down position, the reading should stay constant. The calibration technique was illustrated in Figure 3.12.

Now you are ready to start surveying. Run through the menu to ensure that it is set up correctly for number of points and number of lines, and that you have chosen automatic or manual triggering according to which you want. Enable the logger and clear any grids already in its memory.

English Heritage recommend four readings per metre up the walking line, with lines 1m apart. This is difficult to do with manual logging as the points are so close together. While you are learning to use the instrument, you may find it easier to work to two points per metre for each walking line. You will not lose too much information if you keep your walking lines 1m apart. You can graduate to four readings per metre when you graduate to automatic logging. In automatic mode, there is no difference in difficulty for any number of readings per metre that the mag can cope with, the difficulty is in keeping in pace with the trigger.

It is easiest to start using exactly the same gridding arrangement as the res. This means that you need a team of three: the operator, and someone at each end of the walking line who will move it when you have completed a walk up and back. They should also be dressed non-magnetically. If they cannot manage that, they need to keep at least 2m away from the grid edge until the line is completed. As you complete a line, walk beyond the grid for a few paces. Those moving the walking line can then move behind you while you turn round. It also means that you can be walking steadily before you start the next line.

Do not try to bend down to move the line yourself. The risk of jarring the mag and upsetting the calibration is too great.

a

b

Figure 5.13 There are two ways to hold the mag, either across your body or parallel to it. When you turn round, turn yourself round, but keep the mag pointing the same way. (a) Head north with mag across you, pointing west. (b) Head south, with the mag still pointing west (change hands).

c

d

OR: (c) Head north with the mag pointing north. (d) Head south, with the mag still pointing north.

If you are triggering manually, press the trigger as the end of the mag tube passes over the start mark on the walking line. Note that at one end, this will be one mark in from the base line. Then press the trigger again every time the tube passes over a mark until you reach the end of the line and the mag gives you an end-of-line beep. If you are using automatic triggering, you need to start the trigger as the mag tube passes over the start mark, and then keep walking evenly at the correct pace until it gives an end-of-line beep and stops triggering. You also need to be at the finish point at that time, and it can take some practice to achieve that, especially over rough ground.

If you use automatic, you have to get your pace right and consistent, so you could try a few lines up and back first to check that your pace is correct. To start, try a one second beep and see if you can pace a metre in that time. A metre is a big pace, so you might want to try two half-metre paces in that time. The instrument will be set up initially to beep once a second and to expect you to cover 1m in that time. If that is too fast, check the menu to slow the rate to a speed that you can manage. When you are confident, delete your practice lines and start logging the grid from the beginning. You may be able to speed up once you are experienced.

There are two ways of holding the mag, and you can choose which you prefer. You can either point the head in your main direction or you can carry it across the direction of your walk. These are shown in Figure 5.13. However, once you start one method you must use it consistently. There is another requirement that you do not turn the mag around when you turn around. If the head is pointing forwards on the up line, it must point behind you when you are coming down. This can be awkward as you can neither see nor hear readings being taken. If you carry it across you, you must always keep the head pointing the same way, e.g. west. Then it is pointing to your left on the way up and to your right on the way back. If you are harnessed into the instrument, such as with a Bartington, you obviously cannot do this, and that may degrade results slightly.

If you are confident in your pace, you can dispense with the walking lines and know that you are logging readings at the right separation, but you still need to know where to walk to. You need to use pegs which are clearly visible at 20m to mark your aiming point. Too many pegs can be confusing, as can too little. It is also a good idea to have different types of pegs or flags

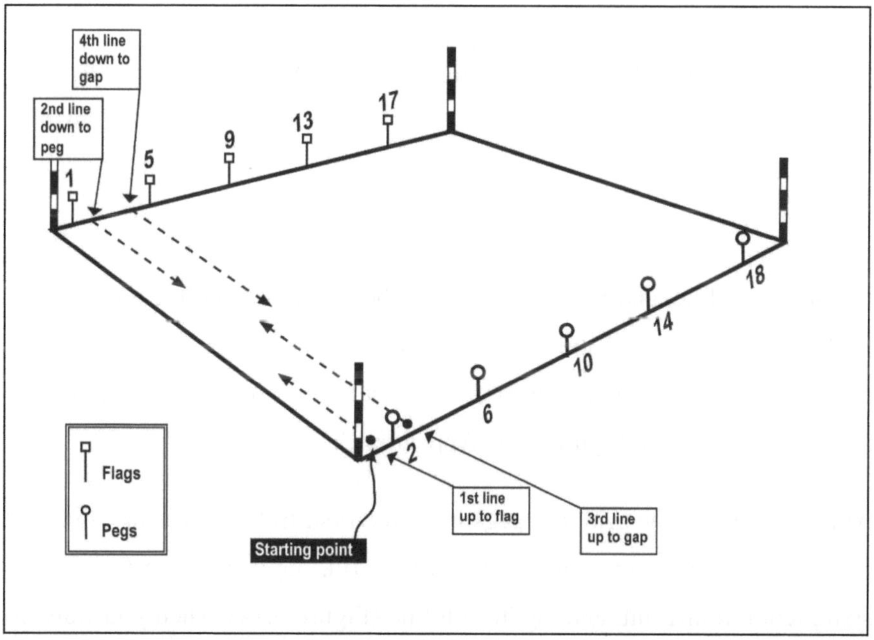

Figure 5.14 For mag without strings, put flags at the top, lines 1, 5, 9, 13, 17 and pegs at the bottom, lines 2, 6, 10, 14, 18.

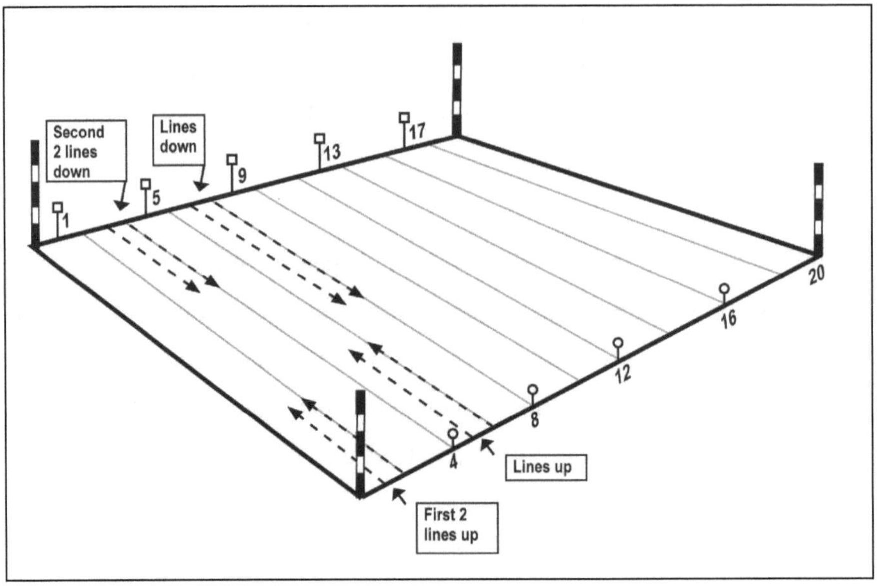

Figure 5.15 For using a double mag, 1m spacing, put flags at the top at 1m, 5m, 9m, 13m, 17m and pegs at the bottom, 4m, 8m, 12m, 16m, (20m).

for each base line, and these pegs need to be different from the corner posts of the grid. The pegs at the south base line need to be very distinctive as it can be very difficult to pick them out against a low winter sun. Types of peg and flag are shown in Figure C1 in Appendix C.

At the top end of the grid, put pegs in on the first, fifth, ninth, thirteenth and seventeenth line. At the bottom end of the grid, put pegs in on the second, sixth, tenth, fourteenth and eighteenth line (Figure 5.14). Then you walk up the first line towards a flag, and walk down the second line towards a peg. You walk up the third line aiming midway between two flags and you walk down the fourth line aiming midway between two pegs. The pattern then repeats, but you must be sure that you are aiming at the correct peg or flag and correct gap. It is very easy to aim at the wrong one, especially if they are at all difficult to see.

If you are using a twin-sensor instrument such as the Bartington, the pegging arrangement has to be slightly different, as you measure two lines at once in a single direction rather than in an up-down traverse. This is shown in Figure 5.15. The flags are still at positions 1, 5, 9, 13 and 17, but the pegs at the bottom line are now at positions 4, 8, 12, 16 and 20, and you aim the left sensor to pass over flag or peg. In this case you always aim at a flag or a peg, there are no gaps. However, it is still a good idea to keep markers 4m apart. Larger gaps or extra pegs are harder to interpret, and that increases the risk of doing the wrong line.

5.6 Download

The process of downloading and processing data has been described in Chapter 4. The important thing is to find a way of downloading safely as soon as possible. It is preferable to do this immediately after the day's work. Sometimes, you might want to download after only a morning session. However, downloading too soon may give a false impression, so it is not a good idea to download each grid. This is also too time-consuming.

You need a safe place to set up a laptop to download. This should be preferably indoors, but can be in the back of a car. However, it must be beyond the reach of rain, and it is best to shelter the computer from direct sunlight so that you can see the screen without difficulty. Ensure your batteries are fully charged. The instrument will be supplied with a download cable to connect it to the laptop. Ensure that the instrument is not in logging mode before you connect to it, or you may accidentally start another grid! You will need to have completed the last grid before the instrument will download. Part-finished grids are not acceptable. If there was only a partial grid on the ground, you will need to pad it out with dummy readings to complete the grid.

Different instruments and different software will have differing download procedures, so exact rules cannot be given here. You will need to ensure that the grid numbers are placed in the right order onto a map, and that the data flow, parallel or zig-zag is set correctly. Note some instruments may be walked in a zig-zag fashion, but may automatically store the data as a parallel pattern. You may want to keep a paper record of grid positions as you go along.

Remember also, that if you have already downloaded data, the first grid in the instrument will not be grid 1 on the map. If you have already put 24 grids into the software, the first grid of the next download will be 25. Then the instrument's grid 2 has to become grid 26, and so on.

Once you have safely downloaded the data and are happy that it has produced the correct picture, you should make a back-up file of your project. You then need to delete the grids from the instrument's memory before doing more survey. Make sure your data set is safe before you delete it from the instrument's memory.

5.7 Contour survey

Adding contours gives an extra dimension to your survey. It can be very important if you have water courses, as it will then be apparent which way water would flow, and this will affect the interpretation. A word of caution, however. The survey will only show what is happening at the surface, not at the level of the features you are detecting. You might expect these to be broadly parallel, but you must remember that if the ground has been ploughed, soil will have moved downslope so, originally, gradients will have been steeper than they appear now.

You will need to choose an origin for your survey. You might want to choose a grid post near the centre of your survey, or to one edge. Depending on your survey instrument, you may have the choice of using national or global coordinates, or using your own local coordinates. If you call your origin [0, 0], you must ensure you can distinguish between positive (east and north) and negative (west and south) numbers. You might find it easier to call your origin, say, [1000, 1000] to keep all numbers positive. The first number is always east–west, the second, north–south. Depending on how you have set up your grid, grid north may or may not be magnetic north. You need to be sure which north you are using.

The third number in the sequence will be the height. This may be height reference to your origin, or may be height above sea level according to your choice of instrument and available height data. You need to be aware of what system you are using, and you need to use it consistently throughout your survey.

The data acquired needs to be entered into some form of spreadsheet, either manually or by data logger download, so that suitable software can be used to produce a contour plot which can be overlaid on your print-out.

Now that we have been through the process of how to set up a survey, we will look briefly in the next chapter at examples of surveys of different size and complexity to give an illustration of what can be done.

6

Examples of Surveys

Your survey will have a particular purpose for you, and this will determine the instruments used and the extent. Survey areas could extend from only a hundred square metres to many hectares, even square kilometres. If you want to understand a previous landscape, you need to survey a large area, but if your interest is in a particular feature and you know exactly where it is, you may only need to survey a very small area to glean the details of interest to you. In this chapter, we present and comment on a few surveys by way of illustration of these types.

6.1 Small-scale resistance survey

This survey was conducted at the east end of an English village church, covering an area only some 20m by 25m. The survey covered both res and mag, but only the res is shown in Figure 6.1 as this contained the main detail.

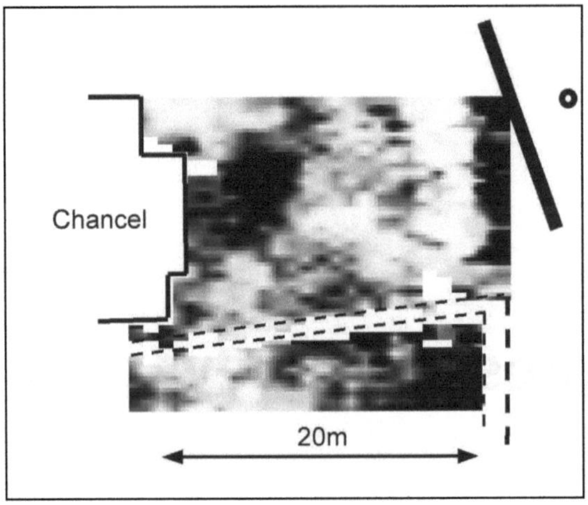

Figure 6.1 A very small area surveyed here showed the goal, the shadow of a building protruding from under a church, but did not show if there was anything else in the churchyard.

The east end of this church was known to contain masonry from the 12th century. The purpose of the survey was to see if there was a possibility of an earlier church on the site. The survey extended from the east wall to the churchyard wall only. The strong rectangular feature jutting out from the church wall suggests that there may well have been an earlier church protruding beyond the east end. In this case, a small targeted survey was sufficient to answer the research question of whether there was an earlier building under the church. However, there may have been other buildings in the churchyard which we still do not know about.

6.2 Magnetometer surveys

Two surveys of moderate area, between two and four hectares, are shown here as examples of where mag has proved valuable for answering different research questions in different parts of the world, and with differing geology.

The first survey comes from North America. It covered 3.3 ha in an area 240m long and up to 140m wide and used a Geoscan FM36. Its purpose was to try to locate the buildings of a late 19th century Indian agency in the western United States. A plan existed of the buildings but their exact location had been lost. The glacial-based soils here were likely to support work with the mag. The buildings could have been timber, but may also have had foundations of cobbles. The mag was considered sufficient to find beam slots of foundation trenches, and indeed proved so without needing res.

The survey shown in Figure 6.2 discovered regular anomalies which indicated the site of the agency, found possible irrigation ditches and irregular anomalies which may be traces of Indian campsites.

Figure 6.3 is an example from a tiny island off the coast of Shetland, north of the British Isles. Foula is only five kilometres north–south by three kilometres wide and its western half is mountainous. This survey was also done with an FM36. Settlement remains, burning and the Old Red Sandstone geology gave favourable conditions for the mag. The area shown had to be downloaded in two separate pieces either side of the burn, and then joined together, together with mapping details, to form the picture shown.

The northern end of the survey was very steep and the western area beyond the burn was very uneven, while the area to the south of the present buildings

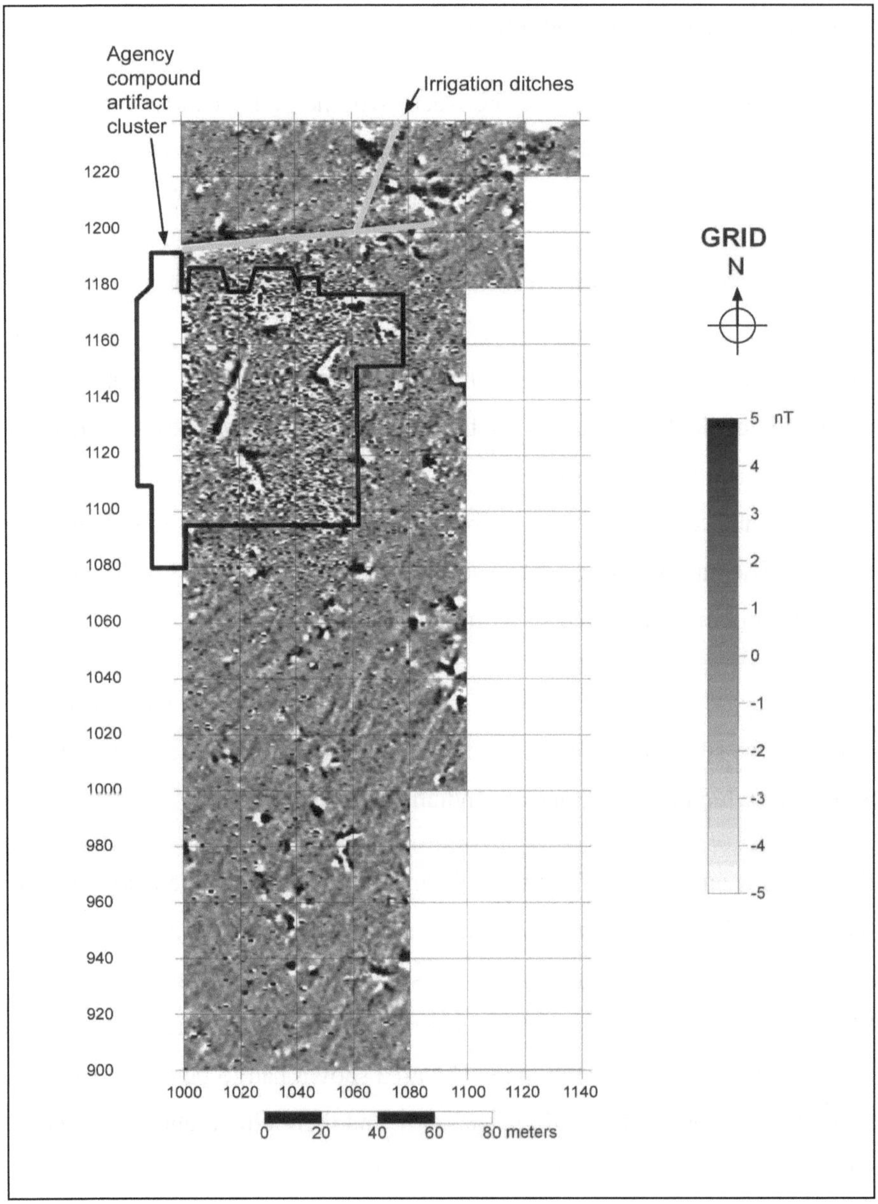

Figure 6.2 The magnetometer survey was sufficiently large to find the Indian agency it was seeking, and also to pick up signs of Indian settlement and irrigation ditches.

Figure 6.3 The survey in Shetland is mounted on a map as it comprised separate surveys east and west of the burn. There is intense activity in the south-east, and plenty of fainter activity elsewhere. The burnt mounds along the burn show strongly.

was flat lush grass. Three strong signals along the burn are burnt mounds, with many pebbles heated by fire then used to heat water. The northern and western areas also show signs of occupation, although these are faint, and a water pipe gives a strong signal towards the south.

The yard immediately north of the buildings shows intense activity, and the flat grounds south-east of the buildings are seen to be an area of settlement no longer visible on the surface. The area of about 4 ha was sufficient to throw light on prehistoric and historic landscapes not previously explored. The area covered was limited by the length of stay on the island rather than by geographical limits.

6.3 Resistance pseudo-section survey

This survey is of Roman remains at the eastern end of the Roman Empire, in Anatolia, north-east Turkey, close to the Black Sea. This was a military centre at Satala, controlling two routes, one heading east into Asia Minor, the other connecting the Black Sea with the eastern Mediterranean Sea. The area was subject to gradiometer survey, which revealed much activity but with limited definition.

Resistance was not measured by a conventional twin-probe device, but by building up a series of pseudo-sections at 2m intervals and combining the results from each at the same depth, as discussed in Section 3.1. This survey covered only about half of the area covered by the gradiometer survey. The result of combining rows of data, each 2m apart, at a given depth, produced an area display, rather like a 'time slice' obtained with ground-penetrating

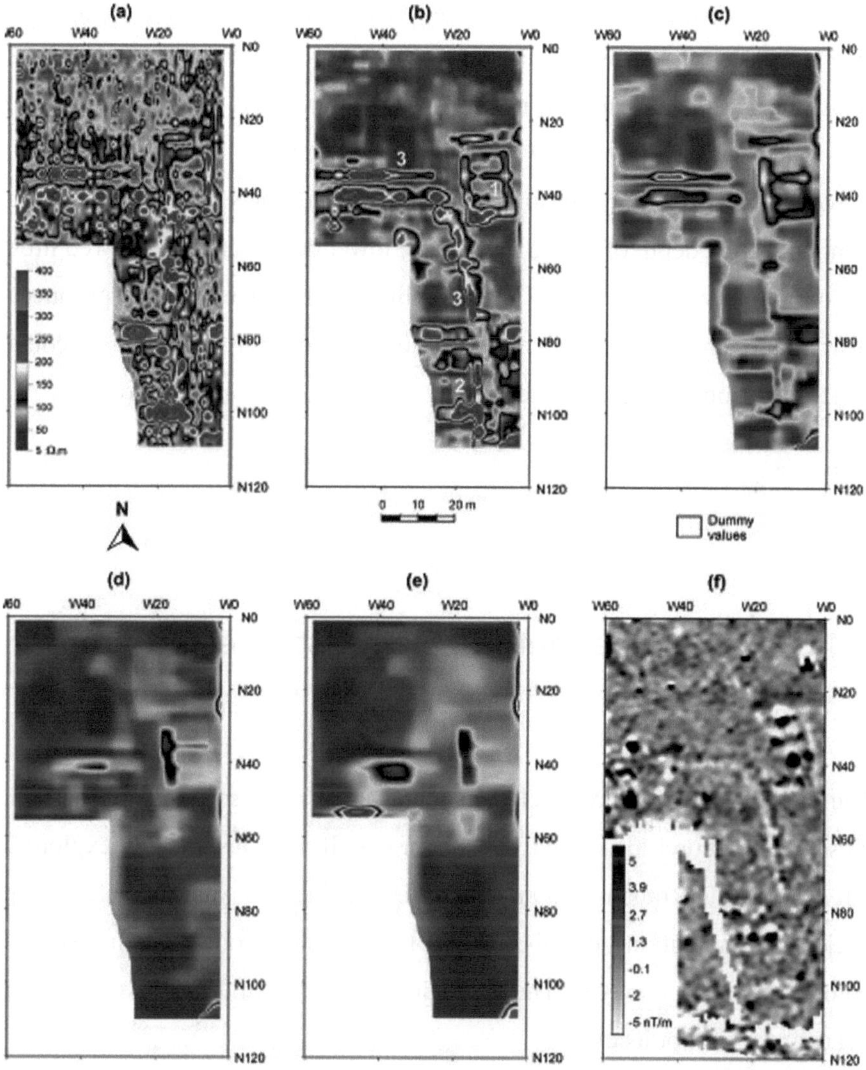

Figure 6.4 The picture bottom right is magnetometry. The other five are depth slices obtained by resistance profiling, and are Roman military structures at Satala, Turkey. Depth (a) is 0.125m, (b) 0.51m, (c) 1.09m, (d) 1.77m and (e) 3.4m. Main detail is in (a) and (b), but there is still significant detail in (c) and vestigial effects in (d) and (e). See also Plate 10.

radar. The plots shown in Figure 6.4 (which is also reproduced as Plate 10 in the colour section) show the presence of masonry structures at three depths below the surface, 0.25m, 0.79m and 1.41m. From this it was clear that there was much activity just below the surface, which would have dominated conventional resistance measurements, but the footings of some buildings went down to a depth of over a metre. The lower depth sections showed little masonry below 1.5m depth. These combined resistance sections produced clear images and were able to add significantly more information than could be gleaned from the gradiometer survey, albeit needing more effort.

6.4 Combined magnetometer, resistance and contour survey

The survey covered three fields, total area 9 ha, in south-west England, close to the church described in Section 6.1. It was designed to explore a 4th century AD Roman villa and its environs. The site was scheduled as an ancient monument and permission had to be gained from English Heritage before it could be surveyed. The approximate position of the villa was evident in advance, but not its extent.

The rock in this area is a bright golden colour, rich in iron, and likely to support work with the mag. It can also be quarried very easily, in sizes that are very useful for building. The res was therefore also likely to produce good results.

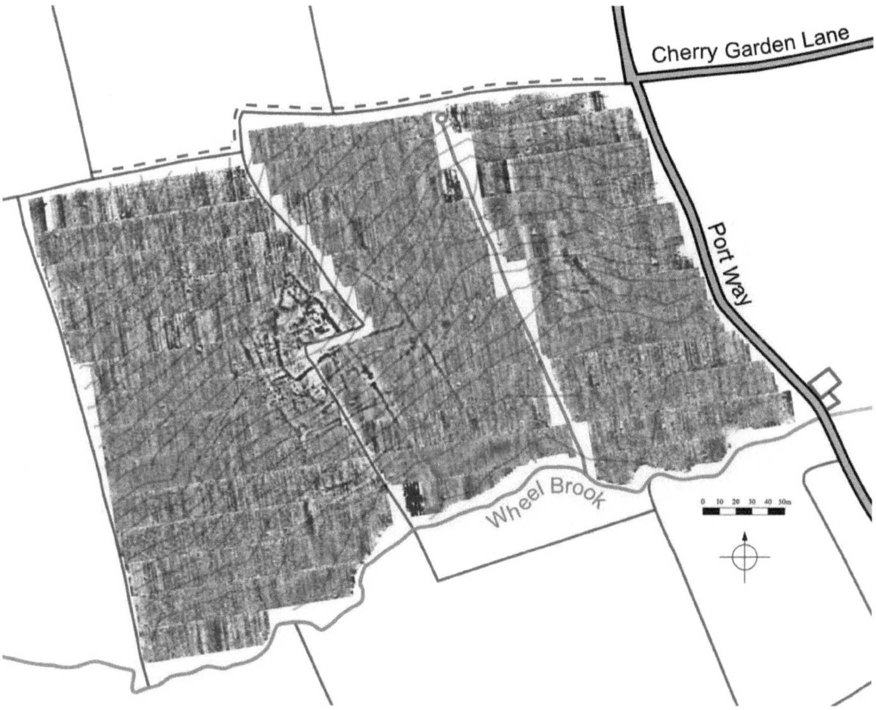

Figure 6.5 Magnetometer survey of three fields, total area 9ha. An enclosure with internal features is clearly visible in the western field near the kink in the hedgeline.

Figure 6.5 shows the mag survey. An area of intense activity within a triangular enclosure shows the villa site just in the westernmost field, with fainter lines extending into the next field. Figure 6.6 is the res survey and shows the main villa range where the mag activity was intense. Now the outline of the building and its individual rooms can be seen. Activity extends south from the building all the way down to the brook. The dark patches in

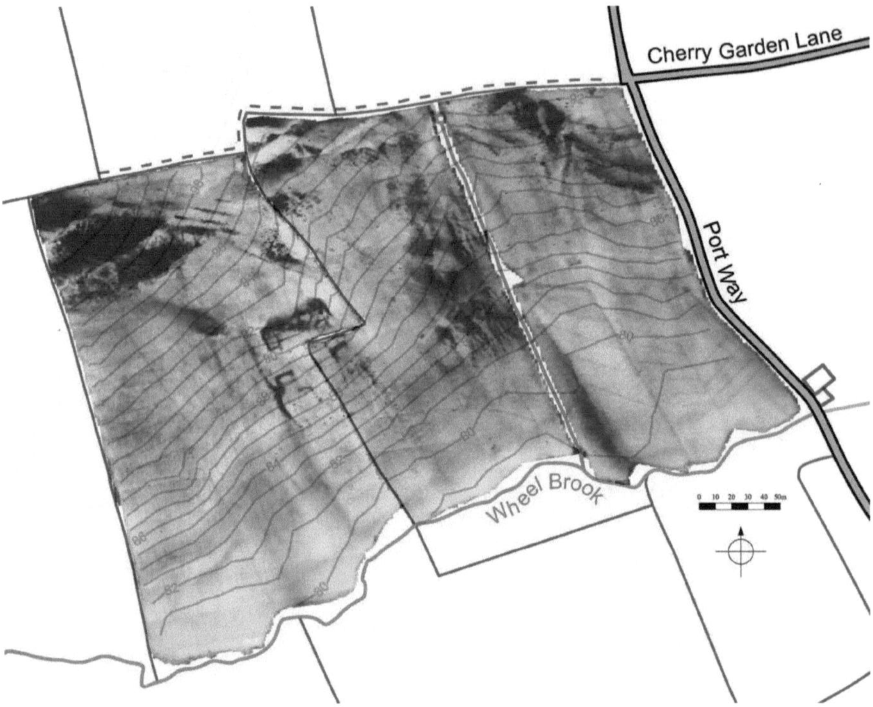

Figure 6.6 Resistance survey of the area. Some modern drainage lines are clearly evident, but a large building can be seen by the kink in the hedgeline, and a range of buildings below it.

the fields are places where the bedrock comes close to the surface. You can also see modern field drains and the dark line where a ditch was filled in with rubble. There was a second building area some 120m southwest of the main range. This was very faint, so not visible in the main picture. The detail is shown separately in Figure 6.7. There are also signs of a farmstead in the extreme east of the fields, where the road kinks.

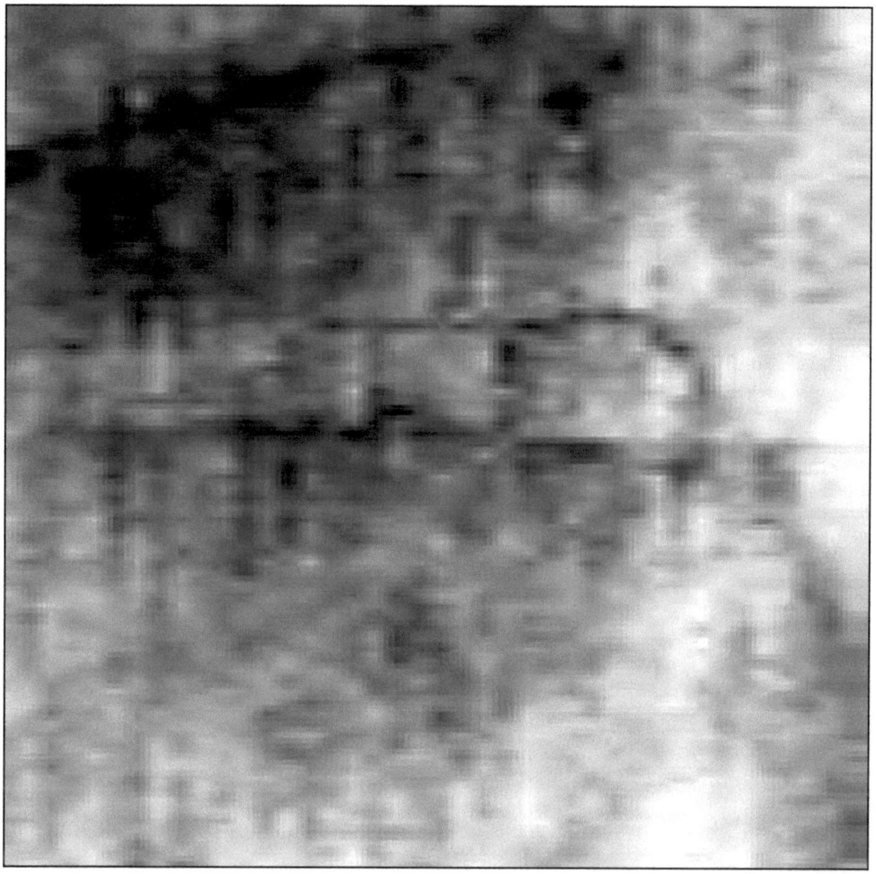

Figure 6.7 Detail of the resistance survey, showing a building some 120m south-west of the main building which is too faint to show on the main plot.

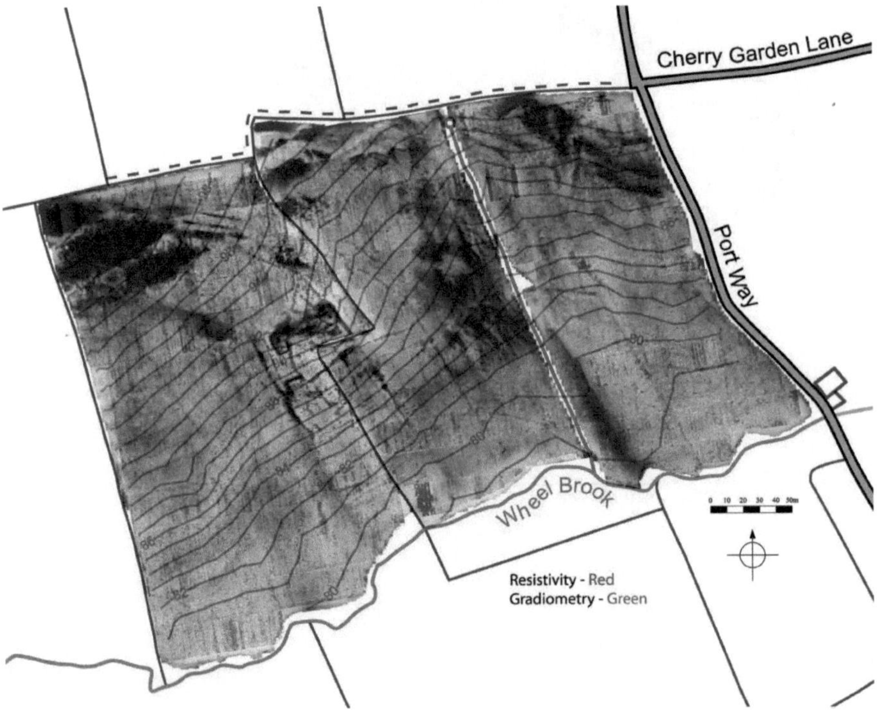

Figure 6.8 Overlay of mag and res (Figures 6.5 and 6.6). Colour is used to distinguish between the instruments (see also plate 11). The large building is now plainly sitting over the ditches of the enclosure.

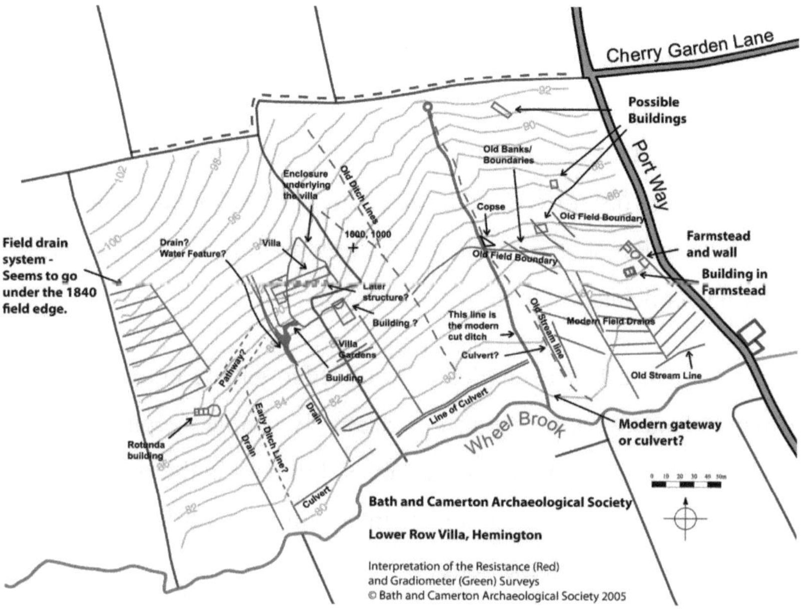

Figure 6.9 An interpretation sketch of the features shown in Figure 6.8 (Plate 11).

Figure 6.8 (also reproduced as Plate 11 in the colour section) shows the mag overlaid on the res, and with contours added. The main villa range can now be seen sitting across the ditch of the triangular enclosure. You wonder if the building started to collapse!! The contours show how the ground slopes steeply from the north down to the brook, while the villa sits on a slight mound. Figure 6.9 (also reproduced as Plate 12 in the colour section) shows an interpretation sketch of all the features found in these fields, including many not associated with the villa.

These are just a few examples of knowledge gained either of buildings or of whole landscapes by the use of relatively simple geophysics equipment. I could easily add many more examples, but now is the time when you can start adding your own! The next chapter will give a brief summary of what we have learnt through the book.

7

To Sum Up

We have now completed our survey of the science and the practicalities needed to undertake geophysics surveys in archaeology, and we have looked at examples of such surveys. Let us recap briefly the main points.

From Chapter 1, you can gather that the first thing you need in doing a survey is confidence, and that comes from knowing how the instruments work and learning how to handle them. However, that is not enough, as there are matters you need to check first, such as whether it is legal and whether the landowner is happy to allow you on his property. I have also tried in this chapter to put the subject in its context, both archaeologically and technically.

In Chapters 2 and 3, we discussed the science of the main geophysics instruments, firstly the pure science and then its application to the main instruments. The principal form of resistance measurement uses a 'twin probe' configuration. This works by injecting a small amount of electrical current into the ground between a probe on the instrument frame and a remote probe and also by measuring the voltage between these points. The resistance is voltage divided by current. A stone under the probes on the frame reduces the amount of current that the ground will take as it has less moisture in it than the surrounding soil. This in turn gives a higher resistance value. If you get a line of high resistance values, you might deduce that you have found a wall.

We also looked at magnetometers and discussed how they were purely passive detectors that sensed the Earth's magnetic field without injecting anything into it. They have to be very sensitive to detect the tiny changes in magnetic field induced by archaeological effects, and this means you have to take great care in handling them, and ensure you are non-magnetic yourself. There are two causes of magnetic signal in the ground. One is burning. The other is the collection of magnetic material in ditches or pits. This magnetic material comes from ashes, iron particles in animal remains, and even magnetic microbes. You could use proton magnetometers, or the more recent Caesium magnetometers, to measure the magnetic field very precisely, but the most common instrument is the fluxgate magnetometer, which measures sufficiently small changes in the magnetic field.

In Chapter 4, we looked at how to get the data from measurements from the instruments into computers, how to arrange the data to give you a picture of what is under the ground, and what is involved in interpreting the results. The latter requires some subjective input. It is not entirely objective and rational.

Chapter 5 concentrated on the practicalities of geophysical survey. First, you must select a suitable site, working either from county-held records or from knowledge of the area. Next, the chapter showed you how to lay out grids in the fields and ensure you can repeat them if you need to go back later. Then there are the practicalities of using the main instruments, how you set them up, how you handle them, how you take care of the downloads. This was followed in Chapter 6 by a few examples of projects ranging in size from 50 square metres to 9 ha.

Finally, I can but wish you good surveying.

Appendix A

Detailed Descriptions of the Main Instruments

A1 RM15 resistance meter

Figure A1 shows the top panel from above. Above that, the front panel (not shown)has a BNC connector (silvery colour), which will drive an old pen plotter, and three main connectors. A six-pin socket connects to a multlplexer for more complex work than we consider here. Next comes a two-pin plug which connects to the long cable which goes into the remote probes. The third connector is also a two-pin plug connecting to the probes on the frame. The second and third connectors are identical, so take care to attach them correctly. The cable coming from the box on the frame has six routes, and you need to put the AD1 adapter cable between them. You then need to use two single leads from the junction box on the bottom of the frame to the

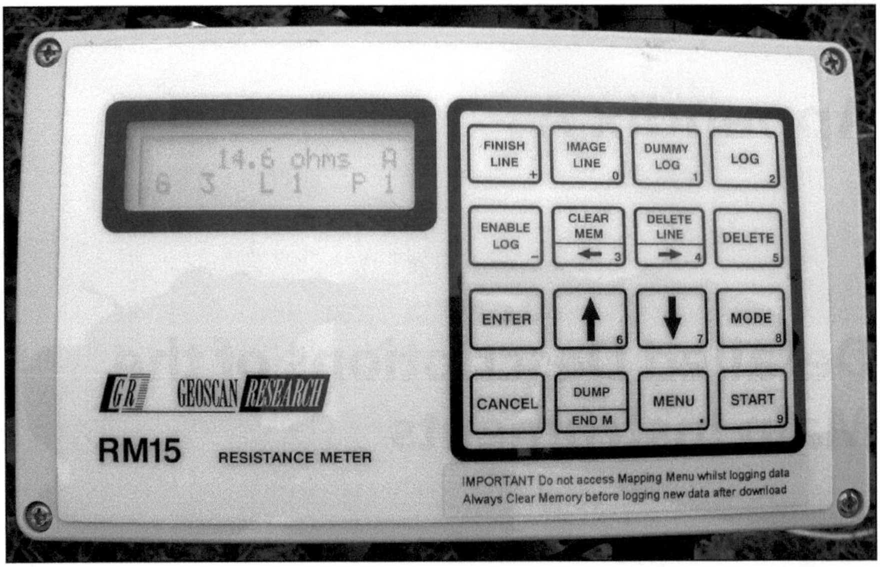

Figure A1 RM15 resistance meter top panel.

probes. Use the shortest lead to connect from the black socket to the probe on the right, then use the next lead to connect the red socket to the left probe.

The back panel (not shown) also has a connector, a six-pin plug. This can take the download connector to a computer or the battery charger connector. It is normal to use rechargeable batteries with the RM15 – you can fit eight ordinary AA batteries but you have to take the lid off to replace them, and this is best avoided. The meter will only charge when switched off. Do not use the charger if you know ordinary batteries are fitted, although very recent models will distinguish between them.

The right-hand side panel contains the on/off switch. Use only the 'off' and 'on' positions, even if the knob can be turned further.

The front panel has the display and 16 buttons. These are not touch buttons, they do need to be pressed gently. The display will give a reading in ohms, or 'open cct' when no reading is being taken. Once you have pressed 'enable log', the lower line of the display will show, for example, for the first grid: G1 L1 P1. Once you have taken that reading (this may flicker, and then give the value logged), press the 'start' button, and the next reading to be taken will be shown: G1 L1 P2. An 'A' will appear to indicate automatic logging and readings will then be taken as you put the probes into the ground. There is no sound to indicate that it has started logging the next reading, but it warbles to say it has completed the reading. It can be a little disconcerting at first, as it appears to hesitate before indicating a logged value. You need to check visually that the display has moved on.

The last reading of the line (if set for 20 readings) will be G1 L1 P20. The device will give a beep rather than a warble to indicate end of line. As you start down the next row, it will read G1 L2 P1, and the reading point will increase to 20 as you go back down the row. At the end of the grid, the device will beep twice.

When you first switch on, the screen will give a value in ohms, for instance '27' (its measurement of resistance at that point), a note like 'PA5' indicating the set up of probes, and a current indicator, usually '1 mA'.

To set up the device, press 'menu' on the bottom line of buttons. This gives you a choice of four menus and, when you press again, a further choice of three menus (see below). You can select which you want by pressing the number of that menu. Look at the buttons carefully and you will see that many of them have numbers as well as names. Once you have finished with that menu, press 'end m', also on the bottom line. This will return you to the list of menus, so you press 'end m' again to get out of menu mode.

Menu 1 is called 'map', with functions:

- grid size (20m);
- sample interval (distance between readings up the line, say 1m);
- traverse interval (distance between lines, 1m);
- traverse mode (zig zag).

You work your way through the menu using the big 'down' arrow. To go back, use the big 'up' arrow. The right arrow on the 'delete line' button lets you increase values, the left arrow on the 'clear mem' button lets you decrease values.

Menu 2 is called 'range', with functions:

- gain (select *1 or *10 to ensure the ohms display reads to the nearest 0.1 ohms for values less than 100 ohms;

- current (select 1 mA unless you need to use 10 mA to get your readings to 0.1 ohms).

Menu 3 is called 'setup' with functions:

• log speed ('medium', unless readings flicker, when you need 'slow' – you can try 'fast', but there is a risk of unstable readings);

• mains frequency (50 Hz UK, 60 Hz US. This filters out any mains effects thoroughly to avoid interference);

• reset RM15? (No).

Menu 4 is called 'array', with functions:

• hardware (use PA5 or PA 1 for simple arrays, do not use UD);

• interface (AD1 for simple twin probe set up).

Menu 5 is called 'comms', with functions:

• baud rate (9600 unless using a very old computer);

• data separator (no space for most download operations).

Menu 6 is called 'progr', with functions:

• program number (1); and

• functions for other programs not used here.

Menu 7 is called 'status', with functions:

• battery voltage (near 10 V, but recharge if it drops near 9V), RM15 version.

Bottom line of buttons

- 'Start' will start logging once 'enable log' has been pressed; thereafter, the device will auto log if set up correctly, until the sequence is interrupted.

- The 'dump' function is used on downloads. The computer will normally tell you to press 'dump' on the RM15 to start the download.

- The 'cancel' button can be used to stop a download midway through, and also if the RM15 says probe resistance is too high.

Second line up

- The big arrows are used for navigating through each menu.

- 'Enter' and 'mode' are only used in multiplex modes, outside our remit here.

Third line up

- 'Enable log' needs to be pressed before the RM15 will start to log readings. It can be pressed again to disable logging. If you have downloaded the grids in memory, you need to clear them (while in 'enable log' mode) before entering any more data.

- 'Clear mem'. Press this and hold it down until it beeps four times, and wait while it says 'clearing memory'. This also has the left button for use in menu mode.

• 'Delete line' will remove the line of data you are entering or have just finished, if you think you have made a mistake. Press it and hold down until it beeps four times. Repeat this if you need to remove more lines of data.

• 'Delete' lets you remove the last reading. Press it and listen for the beep. The last reading will have been removed, ready for you to log again. You will need to log this by hand to get back to auto mode.

Top line

• 'Finish line' – if, say, you can only take the first 12 readings in a line before you hit an obstruction, pressing this will fill in dummy values to the end of the line.

• 'Image line' – if the next row is short as well, and you can only take the same number of readings, turn round after you have pressed 'finish line', and then press 'image line'. This will enter dummies from the start of the next line to the point where you have turned round, and will select the next number in the logging sequence.

• 'Dummy log' lets you put in a single dummy value, for instance, if the next reading is on a concrete path. Press this for the number of points you have to miss in the line.

• 'Log' lets you log readings by hand if for some reason you do not want to use auto log.

The RM15 frame should have a mounting plate with screws protruding through it and the box should sit on the plate. Engage the screws in the threads on the underside of the box and tighten them up.

The frame will normally be fitted with a half-metre transom at the bottom, with the probes in place. There is an option to use a one-metre frame and increase the probe spacing to one metre but this is unwieldy. If you want to use this to penetrate deeper, unscrew the probes and remove the wires connecting them to the junction box. Unclip the plates on the end of the frame and, using the same locating spigots, install the long transom into holes in this transom. Then screw the probes into the end holes of the long transom. If the probes have a square plate, make sure this sits flat under the transom with the washers above. Then use leads of the best length to connect the probes to the black and red sockets on the junction box. Thread the leads through the U-shaped brackets on the transom so that they do not dangle loose and get caught on vegetation. The normal set-up is shown in Figure A2.

A2 TR/CIA resistance meter

Figure A3 shows the top panel of the TR/CIA resistance meter. Above that, the front panel has two connectors. The two-pin connects to the frame. The four-pin can connect either to the remote probes in the field or to the download cable.

The back panel houses two battery holders used for 9V alkali (rectangular) non-rechargeable batteries. When low, they should always be replaced together, not one at a time. Make sure they sit with the wide battery terminal in the wide gap in the housing, and push them firmly home. There is a switch for off and on, and also for HCR, an extra ‘on’ setting which lets the meter take readings as soon as there is any current flow. This is useful on very hard

Figure A2 RM15. Connection of leads and transom to the frame.

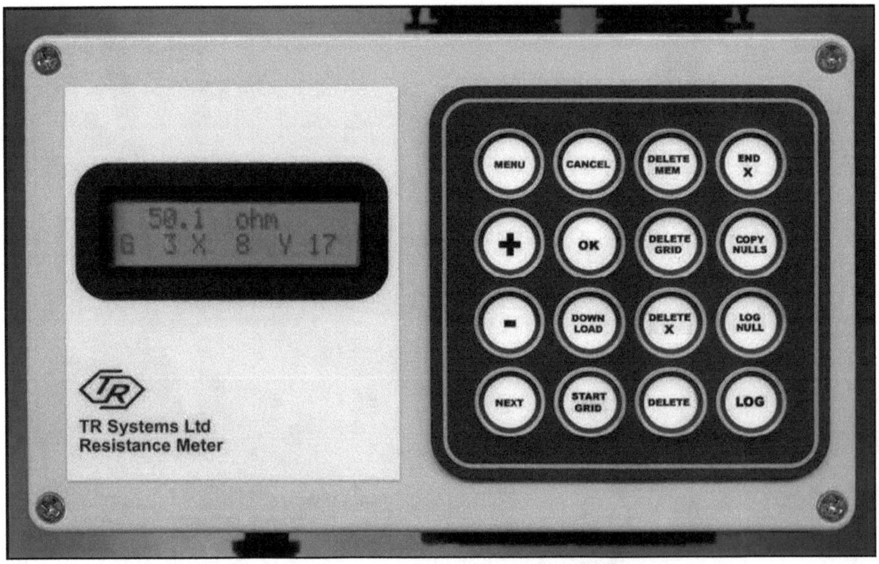

Figure A3 TR/CIA resistance meter top panel.

baked ground, but it can easily lead to wrong values, so it should not be used unless necessary.

The front panel has the display and 16 buttons. These are not touch buttons, they do need to be pressed gently. The display will give a reading in ohms, or '----' when there is no reading to be taken. It will give a BATT warning if batteries need changing and other warnings if the batteries run out completely. It will also give a warning if you pull the probes out of the ground too soon, along with an audible alarm. Once you have pressed 'start grid', the lower line of the display will, for example for the first grid, show G1 X0 Y0. Note that it starts counting from 0, not 1. Once you have taken that reading (this may flicker, and then give the value logged), the next reading will be shown: G1 X0 Y1. The last reading of the line (if set for 20 readings) will be G1 X0 Y19. As you start down the next row, it will read G1 X1 Y19, and the reading point will decrease back down to 0.

Left-hand line: menu controls

• Press 'menu' at top of line. The first indicator will be remaining battery life.

• Then press 'next' at the bottom of the line. This will show you the amount of memory remaining, and also the number of full grids you can do with that.

• Press 'next' again to show the range of ohms. If you can work on the 200 ohm setting, do so. Otherwise press '+' and it will go to 2000 ohms. If you press '–', it will go to 20 ohms, if you are working in very wet conditions.

• Press 'next' again to set filter type. This determines how many readings it averages. The rural setting takes least, and takes 0.5 seconds to get a reading. The semirural takes 1.5 seconds, and the urban 2.5 seconds. Use the rural setting if you can or your progress will be very slow.

• Press 'next' to take you to grid size: number of X means number of lines walked, normally 20. The next position is number of Y, which points along each line, typically 20, but can be higher. X/Y pitch lets you set increments of 1 (1, 2, 3…) or half (0.5, 1, 1.5…).

• Next, logging mode is normally set to 'insert+LCR'. That means it will take a reading when current flows well, not on first touching the ground. By pressing '+' you can go to 'on insert' (equivalent to HCR setting) so you get a reading even with poor contact, but the reading may not be true. Pressing '+' again takes you to an option for manual logging. You have to press the log button every time.

Second row

• 'Cancel'or 'OK' – press one of these to get out of the menu mode.

• 'Download' – with most software, you must press this first before pressing the download button on the computer. The software will normally warn you of this.

• 'Start grid' – press this button when you are ready to go. However, if you have downloaded your last grids, you may want to delete memory first, otherwise the next downloading will duplicate grids already downloaded. Once you have pressed 'start grid', grid, line and point details come up on display.

You then have to press 'log' (row 4) to start the first reading. Note: if you are doing a partial grid, and you need to start part way through, you may have to put in 'log nulls' or even 'end X' (from row 4), until the grid reads the place where you are in the grid, for instance G1 L8 P9. If you do not put these in first, you will not end the grid at the right place. When you have put in any nulls, press 'log' to start the readings.

Third row: delete functions

- 'Delete' just removes one reading, if you have for instance accidentally gone over the end of the line. Press and hold until it beeps. You will then have to press 'log' to restart the measurement sequence.

- 'Delete X' will take out a whole line, for example, if you have got your place on the grid muddled up.

- 'Delete grid' takes out the whole grid, if you have gone seriously wrong. It will give a warning 'delete grids?' but if you hit 'OK' (row 2), it will only remove the current or last grid.

- 'Delete mem' empties the memory, so only use this after you have downloaded, checked your data is good, and backed up the file. It asks you to confirm, but if you hit 'OK', there is no going back. Otherwise, hit 'cancel' to take control again.

Fourth row

- 'Log' is used to start a new grid, or to restart the sequence if you have interrupted with log nulls or deletes.

- 'Log null' lets you put in dummy readings, for instance if you have to cross a path, and need blank readings over the concrete.

- 'Copy nulls' is very useful if you hit a hedge. Press 'copy nulls' and the meter will put in dummies to the end of the row, and an equal number of dummies coming back down the next row, so all you have to do is turn round, put the res in the ground at the same mark on the next line, and press 'log' to restart the measurements.

- 'End X' is another way of completing the line with dummies; for example, if a hedge prevents you from finishing a row.

Downloading

The TR comes with its own download software. This has some processing as well as providing a secure way to get data from res to computer. To download, first press the 'download' button on the res (row 2), then press the download icon (cable) at the top left of the panel. You should then see activity bottom right as the download takes place. On completion, you will get a 'save' panel to select the appropriate folder, and a chance to alter the grid numbers. For instance, if you have already done 15 grids in this folder, you need to alter grid number 1 to 16, 2 to 17 and so on. Downloaded files can be converted to text files so that they can be modified and transferred to other software.

You can open a grid which will let you enter your downloaded data into rectangles or squares in the correct order, but you have to specify the size of the grid to fit the downloads into. You can merge and match the downloaded data and filter or clip the data. Figure A4 shows a typical screen of the downloaded data for processing, with the grids arranged as the area surveyed.

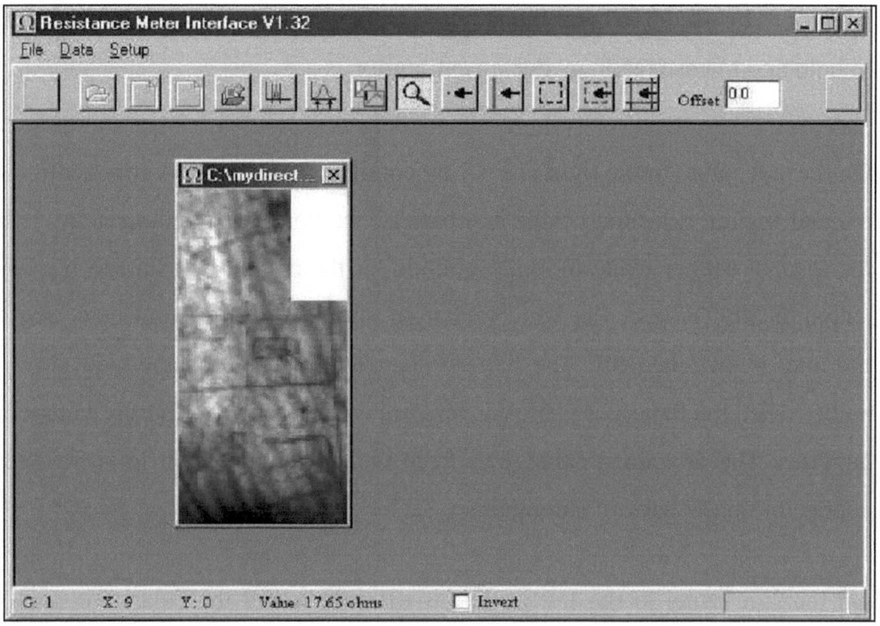

Figure A4 TR/CIA software screen.

A3 FM256

Figure A5 shows the top panel for operating the FM256. The earlier FM36 has a very similar panel, and operates in a very similar way, but lacks some of the functions of the later machine. The extra blank boxes attached to the FM256 contain the main batteries and the memory backup battery. The FM36 has the batteries in the main box. The box next to the handle contains a button which can be used to trigger manually, start or stop automatic triggering, and to hold and release readings during calibration.

On the side panel to the right is a 6-way connector. Three items connect to it: manual trigger, download cable and battery charger. Do not connect any but the manual trigger while in logging mode, or the device will start to trigger automatically. Indeed, the battery charger should only be connected when the mag is switched off. The battery charge light on the same side glows amber when the battery is charging, and red when connected to fully-charged batteries. The download cable goes from the 6-way connector to an RS232 connector on the back of a computer.

On the side panel to the left is the on–off switch and the balance control. The mag should be taken out of its box and switched on 15 minutes before you want to use it, and should be left switched on through the working day in the field. It must be switched off to recharge the batteries. The small hole admits the adjuster (like a small screwdriver) when you do the up–down calibration. You need to feel it engage in the screw head before turning it gently to change the reading to balance the up and down readings during calibration. The adjuster is slightly magnetic, so it must be kept well away from the mag during use.

The top panel has the display and 12 touch buttons. Ten of these have two functions. The main feature of the display is the reading. This may typically say '3.4 nT'. Below it, when you enable log, appears the number of the next reading to be taken, perhaps 'G3 L8 P9'. This means you are about to take point 9 of line 8 in your third grid.

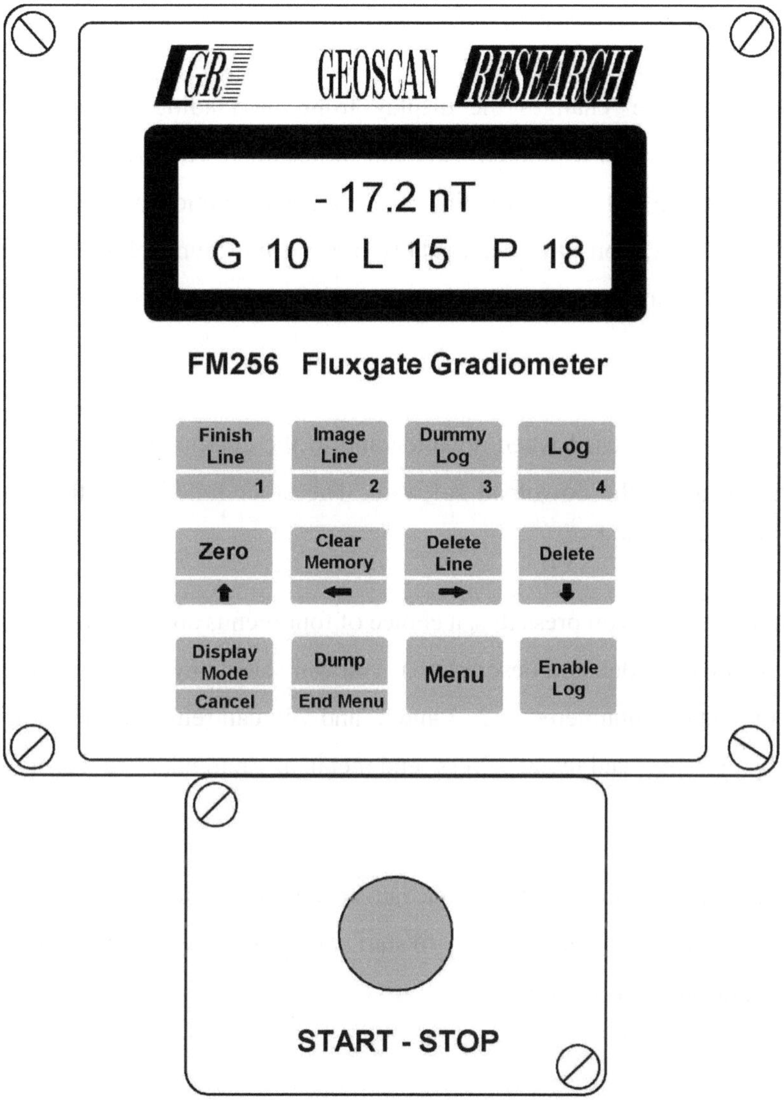

Figure A5 FM256 magnetometer panel.

Bottom line

• 'Display mode' changes the display from the reading, as in '3.4 nT' to a small dial with a pointer. This can be useful if you are just scanning the ground and not taking readings. The second function of this button is 'cancel'. If, say, you have accidentally pressed the 'dump' button when in the field, and have got an error message, you can press 'cancel' to clear your mistake.

• 'Dump' is only used when downloading to the computer, and only when instructed to by the download software. The other function of the 'dump' button is 'end menu'.

• 'Menu' – when you press this, a choice of four menus appears on the screen (see below for a detailed description). You can select any of these using the top row buttons numbered 1, 2, 3 and 4, and you can return to the choice of four by pressing 'end menu'. Press 'end menu' again to take you out of menu mode.

• 'Enable log' switches you in and out of logging mode. Only switch into logging mode when you are ready to start surveying. This reduces the risk of picking up spurious readings.

Menu 1

This menu is for mapping, that is, adjusting the size of your grid. You scroll through in steps, using the 'down' arrow, as follows:

Scroll 1: This lets you choose the length of your grid, ie, the length of your walking rope, perhaps 20m.

Scroll 2: Here you choose the sample interval, ie, how far between the points you take a measurement, for instance 0.5m or 0.25m.

Scroll 3: This gives you grid width (the length of your base line) which is usually the same as the grid length.

Scroll 4: Then comes the traverse interval. This is the separation between your walk up and your walk back, and is typically 1m. (Note the walking lines will be 2m apart, but you walk back in the gap between them.)

Scroll 5: When you scroll on, you can choose 'parallel' or 'zig-zag' mode. It is usual to walk 'zig-zag'; you turn round at the top of the line and come back to the base line still taking readings. In 'parallel' mode, you always have to start from the base line, so once you finish a line you have to walk back to base, so there is a lot more walking to be done.

Scroll 6: The last scroll lets you alter the automatic trigger rate from its standard value of once a second. If you turn this on, you have to 'end menu' to see the adjustments on the screen, and then 'cancel' to get out of that mode.

Menu 2

This menu is called 'range'. The display normally shows, for instance, '3.4 nT'. If you change the resolution up to 1, it would only show '3 nT' and ignore the decimal point. If you are using the dial-like display, you can change the size of its maximum points.

Menu 3

Menu 3 is called 'set up' and has a number of functions, again accessed by scrolling, using the 'down' arrow.

Scroll 1: The first is 'hold reading', which is very useful during calibration. 'H' will appear on the display if it is on. When you now press the button on the small panel, the display will say 'held' and the value shown will not change. When you press the button again, 'held' will disappear and the mag will return to normal measurement. This is very useful for when you are holding the mag upside down during calibration.

Scroll 2: This scroll removes zero, which makes it easier to balance the mag during calibration. For instance, balancing midway between 27.5 and 32.5 is easier than balancing between +1.5 and –3.5.

Scroll 3: The next scroll lets the mag take a number of readings and average them, rather than a single reading.

Scroll 4: lets you choose the number of readings. The more readings averaged, the smoother the plot can become, but the extra readings to average take time, and too many readings to average may slow down your rate of progress.

Scrolls 5 and 6: adjust the visibility of the display. The first switches on a light, which can help on very dull days but drains the batteries. After this, you can adjust the sharpness of the display.

Scroll 7: One scroll takes you to 'log zero drift'. Do not use this unless your software demands it as you have to take an extra reading for each grid as a comparison.

Scroll 8: From this you can choose 'trigger type'. Use 'hand-log' if you are using the manual trigger, or 'internal' in automatic mode, where you have to keep your pace exactly right to finish each line at the right place. The automatic trigger can be started each line with the button on the small box, and will switch off at the end of each line. You can also stop the trigger by pressing the button mid line, for instance to insert blanks.

Scrolls 9 and 10: The next two scrolls, 'baud rate' and 'data format', control the download conditions, and should not be touched without software advice.

Scrolls 11 and 12: 'Scan sound' and 'navigation' switch on the beeps as you go through the menu, and show the scrolling arrows on the screen.

Scroll 13: 'Battery status' does what it says.

Scroll 14: This shows 'set time/date'. If you need to change this, press 'end menu' twice and it shows a set time and date. Use → and ← to correct first the hour, then the ↓ to go on to minutes, again to day, month and year, then press 'cancel' to get back to working mode.

Scrolls 15 and 16: The last two scrolls are for factory use only.

Menu 4: Status

Scroll 1 shows battery voltage. If it drops below 2.3, you need new batteries.

Scroll 2 shows 'time and date'. It is very useful to be able to know the time when you cannot wear your watch.

Scroll 3 gives details of maximum grid numbers (more than you will ever need between downloads)

Scroll 4 shows memory size, also with an option 'view readings' to scroll through the memory to inspect individual readings.

Middle row

Arrows: while you are in menu mode, arrows can also appear in the screen. ↓ (middle row, far right) lets you scroll through the menu you have selected. ↑ (middle row, far left) lets you scroll backwards. → (middle row, centre right) increases the value shown. For instance, if it says '20', → will increase it to '30'. Alternatively, if it says a function is 'off', then → will turn it to 'on'. ← (middle row, centre left) has the opposite effect of reducing the value, or turning 'on' to 'off'.

• 'Zero' lets you set your central value when the mag is calibrated to zero. Press and hold down until it has beeped four times, then hold the mag still until it has finished the zero process.

• 'Clear memory' – press and hold down for four beeps and wait. Only clear memory once you are sure that the download has gone safely and your file is backed up, but do clear memory before you start a new session. It is much easier to remember 'grids 1–14' than 'grids 122–137', and the download is quicker.

• 'Delete line' – if you have messed up a line, for instance by walking too slowly in auto, press 'delete line' and wait for four beeps.

• 'Delete' removes a single reading, useful if you accidentally hit the trigger, or go too near some iron.

Top line

• 'Finish line' and 'image line' are very useful for doing part grids. If you can only get to point 22, say, before you are in the hedge, press 'finish line' to put in blanks up to the end of the row. If you have to turn round at that point and start back, press 'image line' after you have pressed 'finish line'. This will put in the same number of blanks at the start of the next line, so you should get back to your base line at the last reading and have the right number of points and blanks for the software to process the data to give a good grid.

• 'Dummy log' lets you put in individual blanks, for instance if you have to walk over an iron object which would give very high readings. For example, you might have to miss six readings. In this case, you would press 'dummy log' six times to advance to the right place to restart logging.

• 'Log' just lets you log values manually. This can be useful if you are in automatic mode, but need to move more slowly, for instance over a very rough patch of ground.

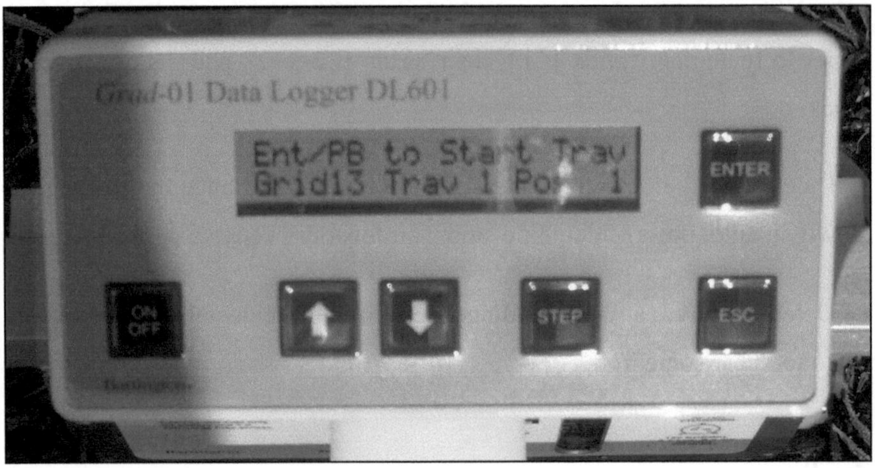

Figure A6 Bartington magnetometer panel.

A4 Bartington 601

The operating panel of the Bartington 601/2 dual gradiometer is shown in Figure A6. This sits in the middle of the crossbar for holding the magnetometer tubes. I describe here the functions of the dual gradiometer. There may be slight differences in the functions for the single gradiometer.

The panel consists of a screen, an on/off switch, two buttons – one to scroll the menu up, the other to scroll down – and a 'step' button which lets you change the value of the selected function. There is an 'enter' and an 'esc' button. The 'enter' and 'esc' functions are duplicated by green and red buttons respectively on the crossbar, sited conveniently for your right thumb.

The back of the panel has two connectors, one for each arm of the gradiometer and an RS232D connector for downloading. There is also a connector from the green and red buttons and a connector from the battery pack. The battery pack also has a connector for the battery charger. The download and battery charger connectors should be kept covered by their caps when in the field.

Menu functions

The screen offers a menu when first switched on. The functions from the top are:

Start survey

Start scan

Output data

Delete data

Set parameters

Adjust gradiometer

System reset

Pressing 'enter' will select the marked function. Scroll to select function. Pressing 'esc' will take you out of that menu to a higher level.

- 'Start survey' – press 'enter' and it will tell you the number of the next grid to do. If you are starting afresh, this should be grid 1. If you press 'enter' again, it will say grid number, then traverse 1, point 1 on the bottom line, 'Enter/PB to start trav'. Press the 'enter' (green) button as you pass your starting point, and you are off. At the end of the first traverse, it will switch off and you have to press green to start the next traverse. As you go up the traverse, it counts the point number and shows the readings to low precision for each of the arms. At the end of the grid, you press green to save the grid and set up the next grid.

- 'Start scan' gives you the precise readings for both left and right arms but does not log the data. This lets you rove around looking for areas of interest. Alternatively, you can use this to find a spot with no signal so that you can set up a calibration position.

'Output data' is for downloading. You first need to be connected to the computer, with the Bartington download software (or Archaeosurveyor) running. The software will give you instructions on when to press the 'enter' button.

- 'Delete data' clears the memory. Use this only after you have downloaded your data and backed it up safely. Do 'delete data' before starting your next survey session after downloading. When you press 'delete data' you get a warning and a chance to press 'esc' before you have to go through with the deletion.

- 'Set parameters' gives you a further menu:

- ‘change pace’ lets you step through walking paces from 0.5m/s (very slow) to 2m/s (very fast) or choose manual logging;

- ‘grid size’ would normally be 20*20, but you could select other sizes: 10*10, 30*30 or 40*40;

• The next menu will usually say ‘start north’, but you can adjust it to other quarters or even mid points. It is for your reference only and does not affect the operation of the device.

You then have the choice of walking a zig-zag pattern or parallel. The data logger automatically sorts the data to parallel even if you walk zig-zag.

You will normally work at 1 line/m, so that you cover two lines a metre apart. You can choose ½ or ¼ metre line spacing, but this involves a complex plan of interleaving your traverses and is best avoided until you are more experienced.

You can choose 1, 2, 4 or 8 samples per metre along traverses. There are some data restrictions on 8. The normal setting is 4, but you will need to use less if you want to use manual logging.

The normal working range is 100 nT, but there is a less sensitive scale if you are tracking iron objects.

You can switch the audio signal on or off. The beep will correspond to a distance walked of one metre at the speed you have set. It also beeps in scan mode, and can be set to beep a warning above a given signal level. There is also a function for loud or quieter beep. The next function allows you to set the level at which the warning occurs.

You can select either one- or two-sensor operations. It is normal to use the two-sensor setting, but if one side breaks down, you can still operate with just one sensor.

You can then set the filter to reject mains electricity interference. This is normally set to 50 Hz in the UK, but would be set to 60 Hz in the US. The final function lets you save the settings, although this can also be done by pressing 'enter' after adjusting any setting.

• 'Adjust gradiometer' allows you to calibrate the device before starting your day's work, but first you have to use scan mode to find a spot where there is little variation in signal (less than 0.5 nT) over an area about two metres square. This operation is similar to that described in detail for the FM256, but the Bartington gives you commands and does the adjusting. To start, face north and press green, then south, then east and then west. While pointing west, you will be told to turn the device upside down to take a reading, then back the normal way up, then back through east, south and north. The Bartington will tell you if the calibration was successful or whether you need to start again.

• 'System reset' – there is a function for system reset, but this is best avoided.

If you need to stop, for instance to put in dummy readings, you press the 'esc' (red) button. This takes you to a menu which offers:

• 'Finish and mirror' – use this if you have reached a hedge midway up a grid and need to put in dummy readings to the end of the traverse and back to the point you stopped at. Then turn round and press green to complete the next traverse.

- Finish grid
- 'Finish traverse' – press this if it is only the one traverse which is blocked.
- 'Change pace' – if your pace is consistently wrong, press 'change speed' and you can step through from 0.5 to 2 m/s to adjust its speed to your walking speed. There is also an option for manual logging using the green button, but it is difficult to do manual logging rapidly.
- 'Back one traverse' – if you are unhappy that you did not get to the end of the traverse at the right time, you can press 'back one traverse', and start it again.
- 'Back one position' – if you have accidentally logged over a bad point (perhaps you noticed a big spike from buried iron), you can 'back one position' to remove that point. Then re-log it, or put in dummy readings.
- 'Enter dummy at (grid position)' – if there are only one or a few points which cannot be taken, then press 'enter dummy' the necessary number of times.
- 'Delete this grid' – if you press 'enter' it asks, 'are you sure?' Enter = delete grid.

The device is quite a cumbersome shape, like a big letter 'H', but it is relatively easy to handle as you wear it. Put on the 'vest' and fasten the straps around your middle and under each arm. The rubber rings on the crossbar then fit on to the hanging hooks. You need the crossbar to be about hip height, so that the bottoms of the tubes are about 30cm above the ground. The magnetometer tubes are stored separately in the box, so you need to unpack each in turn and insert in either end of the crossbar. It is good practice always to keep the same tube on the same side. Install each tube in the splayed end of the crossbar and tighten the nut, making sure that the red arrow points to the front and that each tube is at the same height in the crossbar. Remember to connect the tubes via their connectors under the red arrows.

Full views of the Bartington device are shown in Figures 1.1, 3.13 and Plate 4.

Appendix B

Details of Commonly-used Processing Software

B1Archaeosurveyor

Archaesurveyor has become popular in recent years, and is still being maintained and improved. It runs in Windows, and it can import grids as text files already downloaded to the computer, and also download directly from Geoscan FM36, FM256 magnetometers, Bartington 601/1 and 601/2 single and dual magnetometers, Geoscan RM15 and TR/CIA resistance meters, as well as from a number of other instruments.

The program is produced by DW Consulting, www.dwconsulting.nl. At the time of writing it is at version 2.3.0. It is costly, but a three month trial program can be downloaded.

The program uses a folder for each site, and these can be divided into subfolders for separate instruments, distinct areas or different fields within the site. However, none of the subfolders can be mixed with any of the others.

If you are starting a new project, you have first to create a new folder, as follows.

1. Download to subfolder, including instrument type and grid dimensions.

2. Assemble composite of grids and save it, then close it.

3. Open the grid assembly and save it .

4. Apply processing, which creates a new layer of the composite.

5. View the layer or layers and choose display settings.

6. Save the file or output the file to printer or electronically.

The navigation bar gives you a choice of folders, and of subfolders within a selected folder. Then you have a number of buttons, to choose, for instance 'download'.

Make sure you have highlighted the folder you want to download to. Pressing 'download' will take you through a series of pages. First select the instrument and protocol. It may go automatically to download, but it is best to make sure first that the settings for the instrument are the same as you have used – for instance, the number of points per line you measured may be different from the default setting, and you should check for parallel or zig-zag data (note: this is how the instrument stored it, not necessarily what you walked), direction of your walking relative to grid north and advanced settings for multi-probe/detector instruments (for instance, 2 detectors, spaced 1.0m for a Bartington 601/2).

'Output settings' lets you select grid size in metres, and sampling and traverse intervals. Note that it calls your direction of walk 'x', not the direction of your baseline. For a magnetometer, you may opt to remove drift. Then all grids are centred about zero, so there is no shade change from one grid to the next as the mag calibration drifts slightly. You also select instrument type and units of measurement.

You then select the 'output files', what subfolder you want the grids saved to, and how you want the grids numbered. This can, for instance, start at grid 6 if you have already done 5 grids, or you may have called your first five grids a01 to a05, and wish to call this download b01 to b05. You might also want to put in separate subfolders, say for res and mag. You must enter the number of grids to be downloaded, except for the TR/CIA, where it will find this for you. The next page checks you have the right settings and connections to talk to the instrument.

Going to 'next', the download then starts. This will appear as a series of message boxes, each of which will tell you what to do next. When you have worked through these and seen the download progressing, a message box will tell you when download is complete, and if you OK this, you go on to Grid Assembly.

There will be a box of empty grids visible in the Grid Assembly window, with the downloaded grids sitting in a separate box at the top of the window. North is to the right-hand side of the window, and the downloaded grids have assumed you walked north–south along your lines, starting at southwest corner (top left of each grid). Individual grids or the whole assembly can be transformed to point to other points of the compass or to start at a different corner.

The empty grids form a box of 5 by 5 to start. You can use the buttons on the right to add extra columns or rows to top or bottom to get the empty grid box nearer to the shape that your grid assembly will take.

Drag grids down from the top box to the grid box and see the picture build up. If you get the order wrong, you can drag the grids around to get the right pattern. If you drop a part grid on top of another part grid (perhaps each part taken on separate sides of a stream), the two parts are merged providing there is no overlap. Save the composite when you are happy the grid pattern is right. You then need to save the grid assembly with a recognisable name and close it.

Then open the grid assembly, click on it and then press 'Open Grid Assembly' and then save it. It will now appear as a display that you can process. There are a number of processes you can subject the picture to. The ones you are most likely to use are:

- clip – if you have had a lot of spurious high readings, such as the mag going over a water pipe. Move the sliders to remove those high readings from the display;
- despike, for removing sudden large spikes, perhaps caused when a res probe hits a stone;
- destagger, if your lines have not started at the correct point at each end;
- stretch, if you have walked too fast or too slow and not ended the line at the correct place;

- dedrift, if the instrument's calibration changed over a grid;
- deslope is like a more pronounced dedrift, perhaps when the mag gets too close to a fence;
- destripe evens out each line, setting each line's average value the same, and calculating each point's change from that average;
- edge match makes sure there are no discontinuities in shading from one grid to the next;
- range match ensures that the shading is the same for all values in all grids, so there are no sudden changes at grid edges from black to white. However, this may mask some faint features;
- filters, high pass and low pass, to give the plot a smoother appearance;
- processing produces a new layer of graphics over the unprocessed data.

You can also choose the types of view, such as greyscale shading, colours or contours, as well as 3D views and the range of values covered in extremes of colours or from white to black. This is done in the header designer, which also lets you give the plot a title and add scales for the area and for the range of values. If you have not clipped out spurious high readings, the scale may be too wide, so you will see little detail.

You can choose a number of formats for saving the graphic data so that you can print it or work with it in documents, and a number of ways of designing your page.

You also have facilities for backing up all the data to files either on the main disc or to other devices such as memory sticks.

B2 Geoplot

Geoplot is produced by Geoscan Research, so it is tailored to their instruments – the FM series fluxgate gradiometers and the RM series resistance measurement meters – but it can also process data from other devices. A Bartington gradiometer and TR/CIA resistance meter cannot download directly into it, but grids from them can be downloaded via own-instrument software and imported into Geoplot.

Earlier versions of the program worked with computers set to DOS mode rather than Windows, and this restricted the versatility and usability. The current version 3 works in Windows and it would be better to find this version than use the earlier. These notes refer only to the current version. The program can be obtained from Geoscan Research, and a trial version is available.

First, you need to download or import grids into a directory set up for your particular project. You need to ensure your numbering sequence is correct, but there is a form on which you can work out correct numbers after the first download. You can then edit the grids, for instance removing any large spikes in thc data. When you save a grid after editing, Geoplot will not let you overwrite the raw data, so you have to give it a different name. If you edit grid 5, you might save the edited grid as 5a. All grids for use together must have the same dimensions and same number of data points, and they should all start at the 'bottom left corner'. This would mean, for instance, starting at the western end of a grid and walking your first line heading north.

You then have to make a master grid, on which you fill in on a spreadsheet the numbers of the grids you want to process, placing them in the pattern which you measured them, assuming that your first reading was at the very top left of the top left cell of the sheet. This must then be saved in the directory using a suitable name. You will need separate master grids for res and mag.

Next, you open up a composite, which is the master grid with its grids added. This will give you your first view of all the data. Grid north is to the right, as the arrow points. You can rotate it to north at the top, but it is important not to do that until you have finished all your processing. The panel on the right shows the range and colour range of the readings and a plot of the spread of values. You can use the despike function to remove odd high values, and you can alter the colour or greyscale range.

For res, edge-match is an important process to give continuous range of shades from one grid to the next, providing all your readings were well-matched. Otherwise, effects of geology can produce a very broken appearance to the plot.

For mag, you may need to correct for calibration drift across grids, and you use zero-mean-line to remove striping effects.

You can then apply high pass filters to remove slow changes such as drift or geology or low pass filters to smooth spikes. When you have finished processing, you will then need to save the composite file into the directory.

You will need to back up all three: grids, master grid and composite to ensure your files are safe.

You can then use the 'Publish' function to display your results for printing, adding scales, notes, north arrow and logos.

B3 INSITE

This software package was much used in the 1990s, but has not been maintained so its download functions are now mainly obsolete. It can be used on computers running WINDOWS 98 and later. It requires a parallel (printer) port to hold a deskey (dongle) to license the operator and downloads are via a RS232D serial port. You may still find copies on computers already in use for geophysics, as it is user friendly. For details contact rockside@manx.net.

Its working sequence is as follows:

1. download/import grids from instrument;

2. make grid map;

3. process data;

4. show final display.

There are facilities for backing up the data files to other storage, e.g. memory stick.

The programme opens on to a 'Mission Control' window. Click NEW to start a new project, and fill in the details on the form. All the boxes must be filled, even with provisional information before the software will allow you to go further. You can then choose a name for your SITE, perhaps the name of the field, or the name of the team leader. You can have as many sites as you want, and each can handle four different instruments.

When back at Mission Control, select Data Control and click DO IT. You will most likely have downloaded the instruments by different software to the computer already, so check Import Grids, followed by OK.

The next panel allows you to state your grid dimensions with simple graphics and to select zig-zag or parallel. Note that some instruments may sort the data to parallel even if you walk a zig-zag path. Choose the type of instrument you are using. Use the Browse function to locate your data files, and click on all the ones you want to import, using the control key so you can select several files. OK the selection, OK the data control box and OK on any warning messages. Close the window.

Then select Make Grid Map, DO IT. Click on a square and the first grid will be placed there (you may get a warning message first). Click on other squares in the correct spatial order to place your other grids. If you did not start bottom left heading up, click on the arrow to turn it round or drag and drop it at the corner you started at. Blue arrow is parallel, red arrow is zig-zag data. The Verify function lets you check all data are there by showing lines of actual data. If you have put the grids in the wrong spatial sequence, you can drag and drop them into the right sequence. You can erase grids too: be careful not to leave the erase function on. When finished, click the Return to Mission Control (radio dish) icon.

Select Data Process, DO IT. Your picture will appear small, top left screen. Use the destripe, deblank, de-pike, dewalk, dedrift functions to tidy the picture, pressing DO IT after each. Note that the destripe and dedrift functions cannot always get the best adjustment in auto setting. When you click on Match, the software goes to manual control, and you have to click on the grid you want to process and move the slider to adjust its relative level. The slider has three level settings, changed by double-clicking on the word 'level': blue, small adjustment; green, medium; red, large adjustment. You can also select the other functions and adjust those manually.

When you have the best picture, click Merge, DO IT, select how many points you want to interpolate (half metre, quarter metre or finer). The finer the detail, the bigger the file. Press OK to get your final picture. You can then choose Options to adjust the contrast or to use colours. You can also use shadow plots to accentuate features, but remember to UNDO anything you try before trying another. You can also apply filtering to get a more even picture.

Use Edit -export graphics to store the picture as a graphic, but choose a suitable medium, such as a bitmap.

You can also apply drawing layers, north arrow, scale, range of values by selecting the layer icon (looks like sheets of paper lying flat), but only the graphics layer can be exported. You cannot overlay one method on another for the same site.

Press the Return to Mission Control icon to process further or get to the EXIT button.

Appendix C

Kit List

First collect an instrument, mag or res, and it should come as a complete package, including batteries (either with a charger or with spares), download cable, any special tools to set up and – for res – a frame and a cable with remote probes. However, this will not be enough in itself for you to start a survey. The next item you need is a computer, preferably portable, with download and processing software on it.

There are various ancillary items you need to set up grids. For a start, you need some poles to mark grid corners. These could be dowels or bamboos, but they need to be strong enough to drive into hard ground. It is best if they are 40–50cm long, and they also need to be colourful, either taped or painted in stripes, or fitted with small flags. It can be very difficult to see a thin pole, even on a fine day, so visibility is important. You need at least 20 to set out enough of a grid to work on.

You need at least two tapes to lay out your grids; you could use 30m tapes, but 100m tapes are more versatile. However, they are heavier, and can be difficult to use in strong winds.

Once you have grids, you need to put lines down to mark their top and bottom. These lines should have metre markings on them. Washing line is good, but 20m lengths are too short; you need at least 21m by the time you have made a loop on each end. It is best if they are bright colours, but green does not show up on grass, yellow does not show up on stubble. Coloured insulating tape is the best way of putting on metre markings, but nothing sticks well to washing lines, so you need to check that the marks do not slide around and give you wrong distances. The markers are most visible in red. You need a minimum of four lines, two for the grid you are surveying, and two to set up the next grid.

You also need at least two, preferably three, walking lines, again marked out with metre (or even half metre) spacing. These could also be made from washing line as above, or you can use ropes with marks sewn through so they cannot slip. Rope tends to shrink in the wet, giving poor accuracy, so it is a good idea to shrink any ropes in a warm wash (30°C) before marking them. Again, you will need a 21m length so that you can put loops on the ends.

Lines and ropes will get in a dreadful tangle if they get the chance, so it is a good idea to keep each on a line tidy.

Ropes and lines need pegs to secure them. It is best to use plastic pegs so they cannot interfere with mag. Strong storm pegs can be bought from most camping shops. You need at least 10, preferably 20.

If you are doing mag without walking lines, you will need pegs to mark your sighting points, and it is best to have two distinct sets, one for top line, one for bottom. Extra long tent pegs (30cm) are useful. You may have to cut and mark bamboos or dowels for the other end, but these need to be distinctly different from the grid corner posts. It is essential that they are highly visible, or you will not be able to see them to aim at from 20m away, especially in long grass.

Examples of these items are illustrated in Figure C1 (see also Plate 13 in the colour section).

Figure C1 Ancillary components you need: (top left) grid corner posts; (bottom left) flags and pegs (30cm pegs); (top right), measuring tapes, 30m and 100m shown here; (bottom right), grid lines made from washing line, with pegs (20cm) and wound on a former, and a rope walking line, wound on a former.

You need a compass to find the alignment of your grids. You also need it to set up a calibration point for the mag, so you know where north is. You may also want a hand-held GPS to locate your grid. Five metre accuracy is fine for that, but is not good enough to survey in the grid posts. You may also want a camera to photograph distinctive points so that you can find them again to set up the same grid.

There is also the matter of suitable clothing. This is more important for mag survey, but even with res you need to be fit to survive the weather, hot or cold. You need strong waterproof footware, not sandals. Walking boots can be good for res; if their fixings are non-ferrous, you may be able to use them for mag, but plain rubber Wellington boots without buckles or steel toecaps are safest. Hats are also important, either to keep the sun off you, or to keep warm or to keep the rain off. Gloves also are important; fingerless gloves are an advantage if you have to operate any controls on the equipment.

You may bc ablc to survey wearing thin clothes during the summer but they are not suitable for the rest of the year. You need non-magnetic spectacles if you are doing mag, especially if you have to calibrate it. For mag, you need to check that even trainer bottoms are really non-magnetic and do not have hidden zips, eyelets and studs. Ladies also have to check that they are not wearing wired bras.

For winter wear doing mag you will need a fleece or sweater without zips and a non-magnetic top coat to keep you warm and dry. Most modern heavy weather gear fails the magnetometer test. The best is an old duffle coat, with a plastic poncho to keep off rain. Otherwise buy an old low-cost weatherproof coat so that you can rip off any metal fixings and replace them with Velcro. Overtrousers also need to be free of zips, eyelets and studs.

There are also your own personal needs. You will need to take a drink, preferably a litre, summer or winter, as you can dehydrate significantly in a long day outdoors. You should also take food with you. Travelling offsite to eat will waste a lot of time midday, the best part of the day for surveying. You may want other personal items, such as sun-block, medicines, basic first aid, mobile phone. However, this all adds extra weight to carry around, and you may need to keep some of these well away from the mag.

C1 Sample kit list for a survey project

For res:

- the res box, with good batteries and with frame and moving probes, two remote probes and 50m cable;
- res download cable and computer with good batteries and correct connections and software;
- grid corner posts, minimum 20;
- two measuring tapes, minimum length 30m, 100m preferred;
- four 20m washing lines marked at 1m intervals;
- minimum two lines or ropes marked at 1m or 0.5m intervals for walking;

- minimum 12 plastic tent pegs to secure lines and walking lines;
- sturdy boots;
- suitable clothing and weather protection;
- food and drink to sustain you.

For mag:

- magnetometer with good batteries, with manual logger if appropriate;
- download cable and computer with good batteries and with correct connections and software;
- compass and non-magnetic pegs to mark north–south;
- grid corner posts, minimum 20, preferably 30;
- two measuring tapes, minimum length 30m, 100m preferred.

Either:

- four 20m washing lines marked at 1m intervals;
- minimum two lines or ropes marked at 1m or 0.5m intervals for walking;
- minimum 12 plastic tent pegs to secure lines and walking lines;

or:

- minimum 50 'flags' (short bamboos with tape marking to be visible);
- minimum 50 long (30cm) bright plastic tent pegs.

Also:

- rubber Wellington boots without buckles or any metal;
- non-ferrous clothing (ladies, also check your bras) to provide weather protection;
- receptacle for holding keys, watch, change and metal objects safe while you survey;
- food and drink to sustain you.

Glossary

AC	Alternating current. A form of electricity where the voltage and current continually vary up to a peak and then down to a trough. Mains electricity is like this. It is easier to transmit than direct current.
Amp	Unit of electrical current.
Capacitance	A form of reactance. The effect decreases as frequency increases.
Caesium magnetometer	A device which measures magnetic field by its effect on Caesium atoms. It is akin to a proton magnetometer, but faster and more sensitive. It is costlier and more difficult to set up than a fluxgate magnetometer.
DC	Direct current. A form of electricity where the voltage and current have steady values. Electricity from batteries is like this.
Dongle	A means of protecting software, such that it will only run on particular computers. It has been superceded by later means, but you may well still find dongles on the back of computers. Sometimes called a Deskey.

Download	The process of transferring data from the geophysics instrument to the computer.
Dummy reading	This may sometimes be called a 'nul'. The computer needs full grids of all points when downloading, but you may be unable to do a whole grid, perhaps because it disappears under a hedge, or perhaps because there is some iron that affects the mag. Where you cannot take a reading physically, you put in a dummy reading so that the computer gets the right amount of data for each grid square.
Dumpy level	A simple surveying instrument for measuring the relative height of positions. Now, EDMs and total stations can do this task automatically.
EDM	A surveying instrument which can be used to measure a position very accurately. It has sensors to measure its rotation and tilt, and a laser to measure the distance to that position. It can use these data to measure eastings, northings and relative height.
EH	English Heritage. A statutary body governing the preservation of archaeological sites in England, also holding technical expertise on geophysics techniques. Elsewhere in UK are Historic Scotland, Cadw and Environment and Heritage

	Service in Northern Ireland. There are equivalent organisations in European countries have their equivalents.
Electromagnetic wave	A wave of energy which propagates by changing alternately to magnetic and electric fields. Light is one form of this, radio waves are another.
Fluxgate	A device for detecting minute changes in the Earth's magnetic field, used in magnetometers.
GIS	Geographical Information System. These are computer programs which enable you to build up layers of information, for instance, putting geophysics results on a map, adding contours, or even making three-dimensional models of the landscape.
GPR	Ground-penetrating radar. A device for sending radar waves into the ground, and timing how long they take to reflect off a buried feature. The time can then be converted into depth.
GPS	Global positioning system. A means of locating yourself. That position can be converted to coordinates such as latitude and longitude, as well as height above sea level. It relies on signals from satellites. It is normally accurate to about three metres, but very accurate systems, down to centimetres, can be obtained.

Gradiometer	A magnetic sensing instrument which measures the Earth's magnetic field at two heights above ground level, and subtracts the readings to obtain a difference. This is a common practical form of magnetometer.
HER	Heritage Environment Record (previously known as Sites and Monuments Record – SMR), is a database of archaeological sites kept by each UK county.
Hertz	The unit of frequency, the number of times a wave cycles in a second from peak through trough and back to peak again.
IFA	The Institute of Field Archaeologists. A UK-based organisation for professional archaeologists.
Inductance	A form of electrical reactance. Its effect increases as frequency increases.
Infra-red	Light waves which have wavelengths from about 0.1 millimetre to 1 micrometre, just longer than the band of visible light. We detect these usually as heat. Infra-red lasers can operate at safe low powers and be effective.
ISAP	International Society for Archaeological Prospection. A society based in the University of Bradford which provides a forum for those working in geophysics in archaeology.

Laser	A device which can generate a pulse of highly directed, very ordered light.
Kilo-	A unit multiplied by a thousand.
Magnetic susceptibility	A property of materials which determines their ability to sustain a magnetic field. Measurements of magnetic susceptibility of the ground are done with coil devices and can indicate areas of intense human activity.
Magnetometer	A sensing instrument for measuring the Earth's magnetic field very precisely.
Mega-	A unit multiplied by one million.
Micro-	A unit divided by one million.
Mini-	A unit divided by a thousand.
Nano-	A unit divided by a thousand million.
Ohm (Ω)	Unit of electrical resistance, obtained by dividing potential (voltage) by current (amps).
Prospection	Surveying using geophysics in archaeology, to find new sites.
Proton magnetometer	A type of magnetometer which measures magnetic field by its effect on hydrogen atoms. It was the original type, but is slow and has been superceded.

Radar	A device which sends out a beam of microwaves (a form of electromagnetic radiation) and detects the reflection from objects. The distance to the object can be determined by the time the signal takes to travel.
Reactance	An electrical effect which is similar to resistance, but which can only happen with AC.
Resistance	Electrically, the work you have to do to pass a current with a given voltage. It is measured in ohms.
RS232D	A type of connector used to download data. Many new computers no longer have such a connector, but a lot of instruments still use one. You need to check that computer and instrument use the same connector.
Scheduled monument	An archaeological site protected from interference by law. It is possible to survey these sites, but only under licence from English Heritage (or equivalents for Wales and Scotland).
Tesla	Unit of strength of magnetic field (flux density). 1 Tesla is a very large unit, and we normally work in nanotesla (1 Tesla divided by one thousand million).
Theodolite	A survey instrument which measures the angle and tilt of its telescope very accurately, so that

	you can calculate relative height and bearing. Now, total stations and EDMs do this task automatically.
Timeslice	Used with radar. All the points measured for the same radar return time can be assembled into a plane, which indicates all that the radar detects at a given depth. This enables you to see what features there are at any particular depth. Time here refers to wave travel time, not archaeological time.
Total station	A device which can be operated by one person to determine position very accurately, either using GPS or by acting as an automated EDM.
Twin probe resistance	The commonest way of setting up a geophysics instrument to measure the electrical resistance of the Earth. It has a potential and a current probe on a frame which you carry around, and a potential and a current probe placed at some distance.
Vertical section	Sometimes called a pseudosection. An electrical resistance technique which lets you build up a picture of what is happening archaeologically at different depths underground.
Volt	Unit of electrical potential.

Watt	Unit of power. In electrical terms, it is generally the potential (volts) multiplied by the current (amps).
Wavelength	The distance a wave travels as it cycles from peak through trough and back to peak again. Multiplying the wavelength by the frequency (hertz) gives the speed of the wave.

Further Reading

If your interest in geophysics in archaeology has taken you beyond the scope of this book, the next two I recommend are:

Clark, A, *Seeing beneath the soil*. Routledge, London, 1990, reprinted 2001.

Gaffney, C and Gater, J, *Revealing the buried past.* Tempus, Stroud, 2003.

The first has good scientific explanation, but describes technology now obsolete.

If you are particularly interested in ground penetrating radar, a good website is http://mysite.du.edu/~lconyer/ which leads on to sections on operation and results.

If you are interested in geographical information systems, look at:

Chapman, H, *Landscape archaeology and GIS*. Tempus, Stroud, 2006.

Beyond these are the technical papers, such as those by:

Linford, N, in Reports on Progress in Physics, Institute of Physics Publishing, 2006, Volume 69, pp 2205–2267, *'The application of geophysical methods to archaeological prospection'*.

There are also a number of useful publications on related subjects by English Heritage, available free from their website:

Geophysical survey in archaeological field evaluation. 1995, revised 2008.

Archaeomagnetic dating

Geoarchaeology

Where on Earth are we? The global positioning system (GPS) in archaeological field survey.

A specialist journal is:

Journal of Archaeological Prospection, published by John Wiley and Sons Ltd.

Geophysics papers may also be found in:

Journal of Archaeological Science, published by Elsevier.

Sensing and Imaging, An International Journal, published by Springer.

ISAP, the International Society for Archaeological Prospection, may be contacted through the archaeological department of Bradford University, UK. Their newsletters contain a number of short articles on recent geophysical surveys.

Index

Printing: Mercedes-Druck, Berlin
Binding: Stein + Lehmann, Berlin

Plate 1 Surveying with twin probe resistance measurement. The frame holds a voltage and a current probe, as well as the control box. A cable connects to the remote probes.

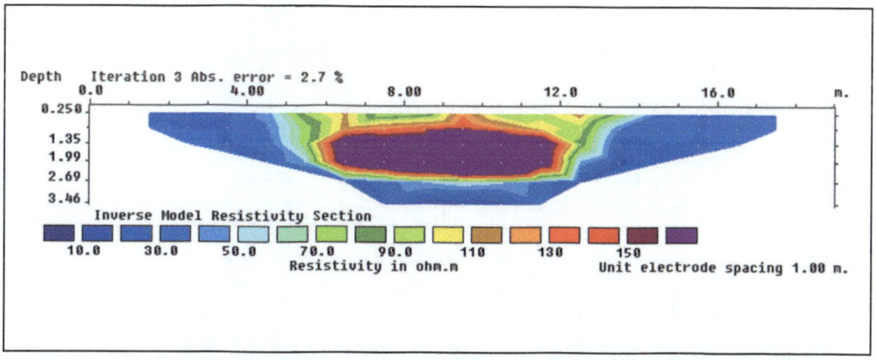

Plate 2 Vertical section obtained by depth profiling. The section gets narrower as less readings can be taken along the line at larger spacings. Blue is low resistance. The red shows a section through a buried building.

Plate 3 The 'wobbly cart' belonging to English Heritage. It is designed to carry caesium magnetometers in a non-magnetic, suspended environment

Plate 4 Surveying using a Bartington 601/2 dual gradiometer. Each tube is an individual gradiometer, so this device can survey two lines at once.

Plate 5 Surveying using a Geoscan FM256 gradiometer. Note the use of a walking string as a guide.

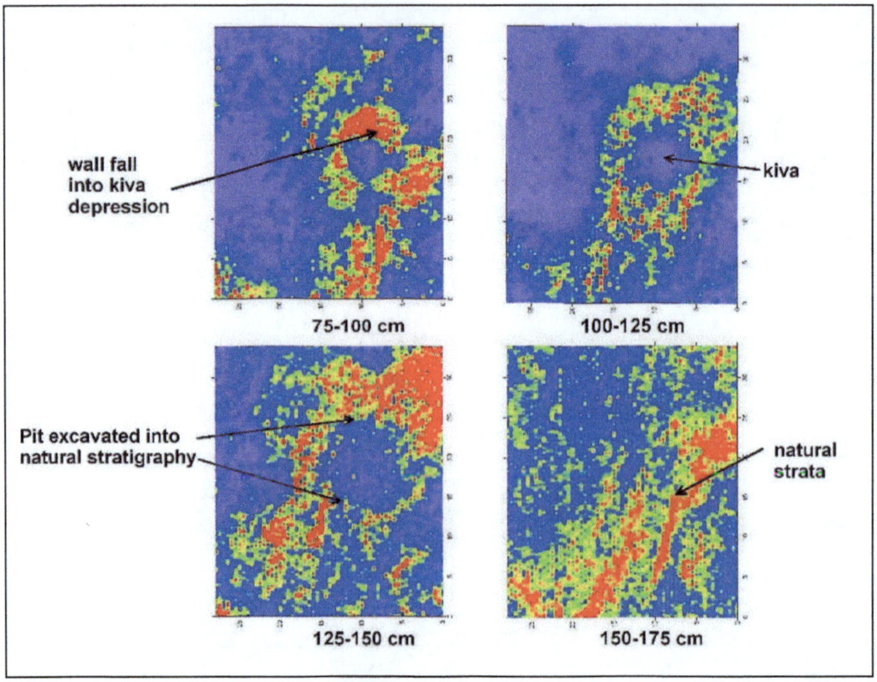

Plate 6 A set of time slices obtained by radar. Each slice represents a different depth under the surface, so you can see at what depth the archaeology is concentrated. Here it is mainly in slices 100–125 and 125–150 cm.

a b

Plate 8 Connectors on a computer (a) connected, (b) disconnected. From the left, USB connector and memory stick, then (other side of mouse connection) RS232D and download cable, then parallel connector and dongle.

Plate 7 A typical radar set ready for use. The controls are on the buggy, with the antenna towed behind, and the wheel behind that measures distance.

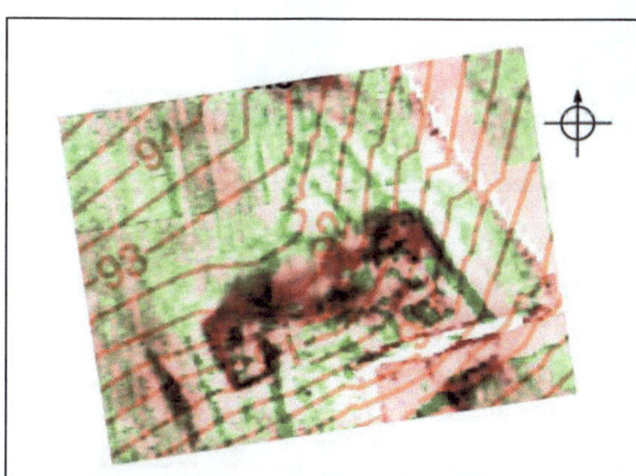

Plate 9 An overlay of a res plot of a Roman villa (red) on a mag plot of a ditched enclosure (green). The building lies right over some ditches. Contours are also overlaid.

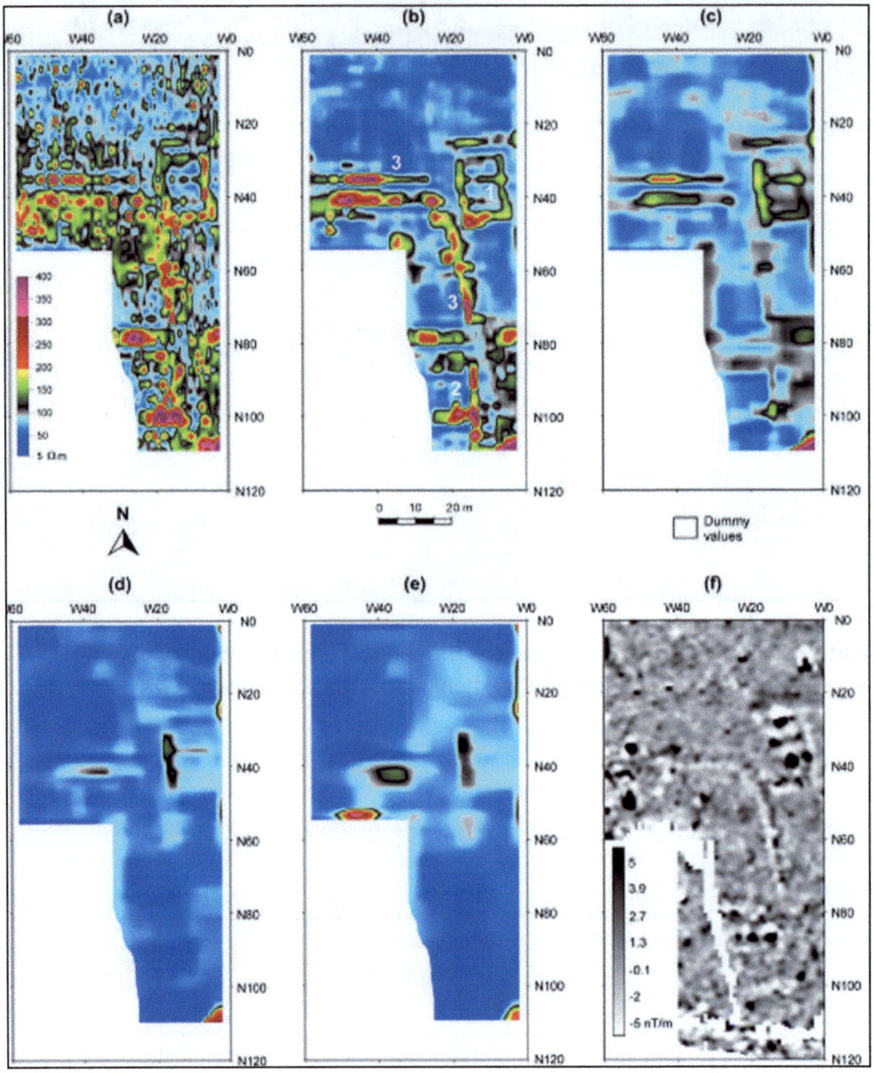

Plate 10 The picture bottom right is magnetometry. The other five are depth slices obtained by resistance profiling, and are Roman military structures at Satala, Turkey. Depth (a) is 0.125m, (b) 0.51m, (c) 1.09m, (d) 1.77m and (e) 3.4m. Main detail is in (a) and (b), but there is still significant detail in (c) and vestigial effects in (d) and (e).

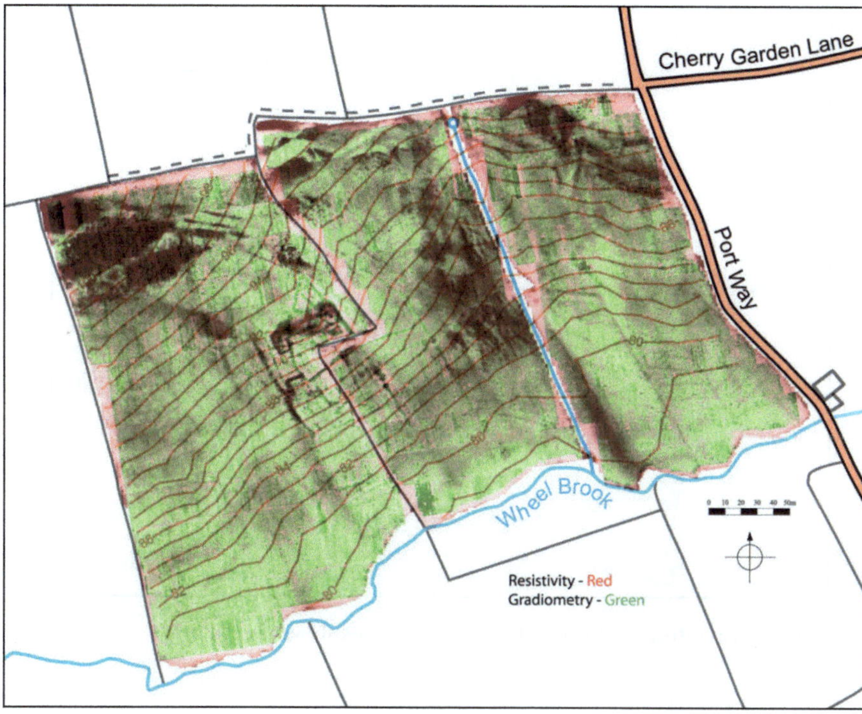

Plate 11 Overlay of mag and res (Figures 6.4 and 6.5). Colour is used to distinguish between the instruments. The large building is now plainly sitting over the ditches of the enclosure.

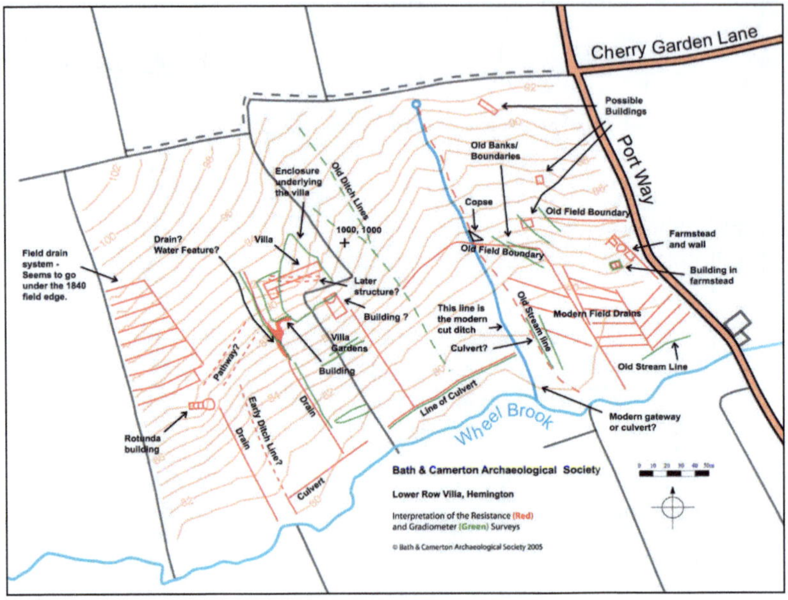

Plate 12 An interpretation sketch of the features shown in Plate 11

Plate 13 Ancilliary components you need: (top left) grid corner posts; (bottom left) flags and pegs (30cm pegs); (top right), measuring tapes, 30m and 100m shown here; (bottom right), grid lines made from washing line, with pegs (20cm) and wound on a former, and a rope walking line, wound on a former.

MIX
Papier aus verantwortungsvollen Quellen
Paper from responsible sources
FSC® C105338

If you have any concerns about our products,
you can contact us on
ProductSafety@springernature.com

In case Publisher is established outside the EU,
the EU authorized representative is:
Springer Nature Customer Service Center GmbH
Europaplatz 3, 69115 Heidelberg, Germany

Printed by Libri Plureos GmbH
in Hamburg, Germany